STEAM ENGINES AT BRESSINGHAM

by the same author

★

MOISTURE GARDENING: HARDY PERENNIALS IN THEIR NATURAL ENVIRONMENT

YOUR BOOK OF TRACTION ENGINES

★

MAKING THE BEST OF ALPINES
(*Floraprint*)

PERENNIALS FOR YOUR GARDEN
(*Floraprint*)

PLANTSMAN'S PROGRESS
(*Dalton*)

PRELUDE TO BRESSINGHAM
(*Dalton*)

BICKERS BROAD (a novel)
(*Breckland Studios*)

STEAM ENGINES AT BRESSINGHAM

The Story of a Live Steam Museum

ALAN BLOOM

FABER AND FABER
London

First published in 1970
This new and revised edition 1976
by Faber and Faber Limited
3 Queen Square London WC1
Printed in Great Britain by
Redwood Burn Limited,
Trowbridge & Esher

ISBN 0 571 10867 9

CONTENTS

ILLUSTRATIONS

INTRODUCTION

This book brings up to date the story of how Bressingham became a live museum for steam engines. Developments since *Steam Engines at Bressingham* was first published in 1970 have been such that to describe them all in detail would have made this new and revised edition pretty lengthy. To avoid this, some cutting of the less relevant passages has been necessary so as to make for more continuity as a narrative.

I could not have foreseen, even when writing the first edition, the extent to which I would be drawn, inextricably, towards further development. I had a compulsion going well beyond the usual urges of a collector who cannot stop collecting. More than anything, there was the unique scope which Bressingham offered to bring these relics of the steam age, with their strong nostalgic pull, back into active retirement. Steam engines seem to engender a strong element of challenge. Engine drivers know this, for even if all is well, the challenge is to keep it so, which calls for concentrated effort. It is, as I know only too well, a challenge to take over a derelict engine and to restore it to life, and the moment it does come back to life brings a rare thrill. But perhaps the greatest stimulant for me, as a collector, was the interest shown by an ever-increasing number of visitors who came to see the engines. This proved not only that we were undertaking a worthwhile task, but that, as a live steam museum, Bressingham could become a self-supporting but non-commercial venture worth every effort to ensure that it would remain so beyond my lifetime. In this belief I quickly shed any ambition for personal ownership. It had to be a case of sharing—with actual owners allowing permanent loan as is the case with all the larger locomotives, and of course with visitors—while those of us here acted as custodians of what is no less than a national heritage of historic and educational value.

I am not an engineer, but a gardener and farmer by profession. Like hundreds of thousands and maybe millions in various walks of life, I have been aware ever since early childhood of a love for steam engines, mingled with excitement and awe. Scarcely anyone, young or old, would have believed that the steam age would come to a rather abrupt end after 1945. Even if steam on the roads and farms was on the way out at that time, on the railways it seemed certain to remain as the main source of power. Yet by the late 1960s British Railways had gone over entirely to diesel and electric power—in what seems now a hasty and ill-considered rush—instead of gradually phasing out steam as did most European countries. Perhaps it was this relatively rapid switch which brought steam nostalgia so quickly into the British mentality, though there is ample evidence that it is now on the increase abroad as well.

As one who has had the good fortune to indulge to the full my own love of steam and to share it with the thousands of others who have not had such an opportunity, I hope this book will have evocative as well as readable qualities. There is, for those of us at Bressingham on whom the task of restoration and maintenance falls, a pretty heavy responsibility. We are not in any position to play trains for our own and public amusement, and must always aim as close to perfection as possible in such matters as restoration, maintenance and safety, so that sometimes pleasure and satisfaction are offset by anxieties. The ultimate goal of having every engine in first-class running condition has so far proved an elusive 'will-o'-the-wisp'. The amount of work still to be accomplished remains formidable, not only because of the need to make additions to the collections whilst they are still available, but through wear, tear and the inevitable deterioration of metal over the years. We have been conscious too, of the need for youngsters to take our place as time takes its toll on us—youngsters who must be taught fast-dying skills connected with steam. Mindful of the vast amount of work to be done, for the foreseeable future we must keep to restricted opening to the public. This is given below, and only in exceptional circumstances, by appointment, can we consider visits at other times and days.

TIMES OF OPENING

Every Sunday from second in May to last in September, inclusive, from 1.30–6.00 p.m.

Every Wednesday during August, from 1.30–5.30 p.m.

Every Thursday from late in May to mid-September, inclusive, from 1.30–5.30 p.m.
Also Easter Sunday and Monday and Bank Holiday Mondays, from 1.30 p.m.

ACKNOWLEDGEMENTS

When, early in 1961, I decided to indulge in traction engines once more by way of a hobby, I had no thought there would ever be more than one. Flora could certainly not have guessed when she encouraged me to acquire 'Bertha', what this would spark off, for when the steam bug became much more than a hobby, she more than anyone, felt its sting. The acknowledgement I owe to her can scarcely find expression in print, but in a different category there comes to mind a succession of helpers on the practical side without whose skill and interest there would in fact have been only one engine at Bressingham.

The first of these was Jack Clements, who first made up for my own lack of skill and knowledge, and at 84 he is still with us. So is Roger Garnham who joined us in 1962 as a fitter, and Don Hubbard who, having been on the Nursery since 1947, came over to steam in 1968. His patience and industry, on exterior restoration work will ensure long life to some of the larger locomotives, and George Garrards' skill as an occasional welder has solved many a problem over the years.

In more recent years Geoff Sands has proved that his qualifications as a locomotive engineer were indeed backed up by practical experience, and what he doesn't know about railway engines is scarcely worth knowing. Each one of our little band of full-time and part-time helpers, must be included, for all are motivated more for love than for money. Most of them mentioned by name in my first book are still active from a distance on our behalf. Since 1970, more such names could be added. All of them, who I like to think of as 'Friends of Bressingham' because of their ever ready willingness to foster its welfare, must be included collectively. They range from those who show a special interest as visitors to those always on the look out for ways of enhancing the project, to voluntary helpers who come, often from a distance, to help on open days, including past and present B.R. drivers, and to those whose interest lies mainly in the locomotives they have here on permanent loan.

And now that at last the Bressingham Steam Museum is registered as a Charitable Trust, this is the place to give the names of those on the Board of Directors: Sir David Follett, Flora E. Bloom, David H. Ward, J. M. Price, Robert Bloom, John Dain, Robert Manning and Mrs. E. M. Fox (Secretary). As Chairman I cannot but feel content, knowing that the future well-being and continuity is safely shared.

I

Expression in Steam

'Who owns this place, mate?' The question came from a coach driver standing at my elbow as I peered at Gwynedd's gauge glass. Until her injector was singing nicely and I'd checked her fire, I made no answer. It doesn't do to neglect either fire or water on a little locomotive, eighty-five years old, if she is having to work fairly hard. There was quite a crowd of visitors building up, and I guessed my questioner had sauntered over, after his passengers had left his coach, to where Gwynedd stood at the platform of the two-foot Nursery Railway.

'I say!—who owns this place? Is he a lord or a millionaire or something?' I took my time before replying, but now as I looked into the coach driver's face, I saw he was serious. 'You've not been here before?' I countered. 'Why, no,' he said. 'Never even heard of the place. I just brought a party from Luton. I took it Bressingham Hall was one of these 'ere stately 'ouses as they call 'em, but this looks different.'

I grinned beneath the sweaty grime one soon acquires from close contact with coal smoke. 'Well,' I said, 'the owner just happens to like steam engines, though he's a nurseryman and farmer by trade.'

'Well, I'll be damned. I can't think why I've never heard of the place before. I s'pose you've worked here a long time?'

'Over twenty years,' I told him, as I turned off the injector before stepping back on the low footplate and blowing the whistle for another journey round the nursery.

I was glad to be working the train alone. I like driving and prefer to do my own stoking, which only amounts to a few shovelfuls of coal for the mile-and-a-half trip, fed if possible at whichever point seems to ensure that the resultant smoke will not drift back over the passengers in the open toast-rack trucks.

Musing over the coach driver, I did not hurry—six or seven miles an hour is quite fast enough for coaches with wooden seats and no springs. Coasting down from the platform under the chestnut trees, passing the new museum shed on one side and the two-acre pond on the other, needed no steam but instead a little check with the handbrake, before the valley bottom was reached. Here the soil was black and peaty, but the track, laid only a year ago, had not sunk so much as I'd feared. The four coaches, home-made bodies on second-hand quarry bogies, rode steadily enough behind, and the passengers were obviously too intent on the long rows of flowering plants they were passing to bother about the lack of springs. The rushy fen ditches and the woods beyond were cool and green.

Gwynedd's gentle rhythmic puffs brought us into the marshy Roydon Fen pastures, where only a few years ago nothing grew but reeds and bushes. Curving back through a coppice of silver birches brought a blaze of colour into closer view. Last winter I'd spent a few enjoyable weeks preparing the formation and laying sleepers for this new extension of half a mile or more, with thoughts then of what it should look like now. Six acres of phlox were growing here on the low sandy mound of 'Freezen Hills' and ten years ago this had been a rabbit-ridden waste of gorse and bramble. This was the farthest point from home, and doubling back with a sight over other colourful fields of nursery stock, I stoked Gwynedd again, ready for the 1-in-40 gradient which gave passengers the sound of sharply powerful puffs and the sight of plenty of smoke till the higher level was reached again.

More smoke lay ahead, from other locomotives or the traction engines in steam at the yard, and here too Flora's Gallopers were merrily revolving to such tunes as 'Colonel Bogey' or 'Farmer's Boy' on the organ. The new car park was filling up nicely as the train ran between it and the alpine plunge beds with their 600,000 potted plants. On these trips, there were too many demands on my attention to indulge in musings as to how all this came about. There was time only for brief mental flashbacks, because uppermost was the prospect of still more exciting developments due for fulfilment before the damp and often dismal summer of 1968 was over.

To think right back had to be a deliberate act; it not only brought relief, it also helped me to get the present and future into perspective. Sometimes, especially during the increasingly frequent nights when sleep was elusive, I had to try and stand outside myself, because steam, which I'd taken up as a hobby, was becoming so much more than a hobby—almost an obsession. I was in danger of becoming blinded by it, of not seeing clearly either my other responsibilities or my own human limitations. But for all that, this indulgence of my love of steam engines had been so rewarding, and the thousands who came here to see them were in themselves some justification. To begin with I could identify myself with any one of these thousands. Steam had its appeal for me, to begin with, on no different level than for those who came to see my engines, with no prospect of owning a single one. Yet here I was with more than a score of engines under my charge, and others due to come soon. My ambition to operate a comprehensive live steam museum had emerged not so long since, and already it was pretty close to fulfilment.

No doubt other visitors must imagine, as had the coach driver, that I was 'a millionaire or something'. We had been up against such an assumption of wealth even from local people. Others, I fancy, thought of me as an eccentric, with steam engines my particular craze. They were entitled to think what they liked, and I much preferred this to the millionaire tab, which was so very far from the truth. In the real sense I was wealthy. Life was very full and very sweet. I'd known hard times and could enjoy a measure of prosperity all the more for that. I have always loved flowers and plants and for years had worked hard to build up a business as a producer of plants. Then when the chance arrived, though rather belatedly, I'd spent my personal earnings up to the hilt in building up a collection of steam engines, which had proved so unexpectedly to marry in well with my love of flowers. A man lucky enough to have the scope for giving full expression to two such interesting subjects as flowers and steam engines must be wealthy, in the true sense of the word. Financial assets, in this context, must take second place. If the thought and energy with which I'd backed my love of plants and steam engines had been channelled into making money, I'm pretty certain that kind of wealth would have come, but I'm equally certain it would have brought little by way of joy and satisfaction for me.

This is not to say that I planned my life deliberately on these lines; nor to deny that if I'd not made mistakes or taken wrong decisions, I might have achieved wealth in both senses. But I could have no complaint on that score. I alone knew how close

I'd been once or twice to the brink of ruin, and had I stumbled over, it would have ruled out wealth for me in either sense. To have narrowly avoided the pitfalls that lie in the path of the headstrong was my good fortune, but I could at least say that none of my successes had been gained without effort. And I'd enjoyed the effort as much as the rewards it brought.

2

A Boy Stares at Steamers

Enthusiasts usually become tongue-tied when asked to explain just what it is about steam that gets them. There is no ready answer. If you try to explain to someone unaffected by it, you tend to give up halfway because you see no hint of comprehension or sympathy in the eyes of the questioner. Some press reporters especially are like that. More than once when I've been asked why I collect engines, I have answered bluntly that there's more satisfaction for me in collecting steam engines than there would be in collecting postage stamps or pottery. There is, of course, much more to it than this and there are different kinds of enthusiasts for steam. Some are interested only in road, stationary or marine engines, others only in railway locomotives. You find men whose interest lies in the purely mechanical aspects, others with a vast store of historical knowledge and able to tell you which design of locomotive hauled the Irish Mail in 1880, and what the performance was. Then there are those whose outlet lies in models, whose imagination is so vivid that a midget layout in a shed with toy trains will remain a joy to the end of their days.

My own interest does not come into any of these categories, nor have I become simply a collector. I have precious little knowledge and even less skill as an engineer, and could no more make a model than fly. I can only admire those who possess such skill and knowledge, and it would be quite vain and useless to envy them. My own categorical definition would, I freely admit, be that of a man who from earliest childhood has been fascinated by steam engines of all kinds and who, on approaching middle age, was offered and took a rare opportunity of giving full expression to this enthusiasm. 'Full expression' means much more than acquiring engines as collector's pieces, antique though most of them are. It means their restoration and preservation as objects capable of coming to life in no more time than it takes for a fire lit in their bellies to turn water into steam. An engine, however beautifully preserved, has little appeal for me if it cannot come to life. And for me full savour only comes when others too can see what happens, when engines awaken to life in steam.

I cannot remember the first steam engine I ever saw, but whether it was one of the traction engines which served the local farmers for threshing, or a railway engine, it made me want to see the next. I was born at Over, a village on the edge of the Cambridgeshire Fens. Swavesey station on the Great Eastern Cambridge–St. Ives line was more than a mile away, and that was where the nursemaid was most often persuaded to take my younger brother and myself in our push-chair, for afternoon outings. She was a kindly soul, but if after a long vain wait for the sight of a train she refused to go up to the signal box and ask if one was expected, down went my regard for her. Towards Cambridge the line was dead straight and passed under a road bridge a mile away. A train could be spotted well beyond that bridge, if indeed one was approaching, but hopeful imagination could play tricks with eyesight and too often we turned homeward both disappointed and resentful. What George's feelings were I do not know. He was a year younger and a much more placid child than I was.

Quite often my luck was in, and though in those years just before the First War it was not a busy line, we might see two or three trains in the space of an hour. Mostly they were goods trains with plain J. 16's or 17's slowly hauling a long line of coal trucks. With inside cylinders, these engines had extension rods that moved in and out with each stroke, just visible beneath the buffer beam. They fascinated me deeply, and as I was prone to nightmares, it was with fair regularity that I dreamed of being fixed to the rails with no escape from an oncoming engine with its inexorable beat and rhythm coming closer and closer till I woke in terror. Perhaps that is one reason why a locomotive still generates a feeling of awe within me.

In those days, and for years afterwards, the Midland Railway had running rights over this Great Eastern line from St. Ives to Cambridge but only one of the four daily trains was allowed to stop at the intermediate stations of Swavesey, Longstanton, Oakington and Histon. The other three ran as expresses and a fine sight they were, even though they carried three coaches at most. Hauled almost invariably by quite small 2-4-0's their red paint shone, and both engines and coaches were spotless. I learned later that these were of Johnson's design, and when I could read saw that some were built at Derby works as long ago as 1879. A few had inside frames, and though these were specially attractive, all these Midland expresses were a fine sight. They were more than this, for local folk who used to tell the time more accurately by their distinctive whistle—a long low note that rose to a higher pitch at the end—than by the chimes of Over church clock. 'Harpas three—there goes the Midland' someone in Over would always say when the wind lay south or west, even though it was out of sight, a mile away. Their speed and reliability were made use of in a very different way, once if not twice in my early years, when some poor harassed soul put an end to his life under those gleaming wheels.

The Great Eastern Railway was held in no great esteem in those parts and in those days. Its passenger trains were not so reliable, nor was its rolling stock so clean and comfortable. The Holden 2-4-0's were fussy little engines though efficient enough to haul half a dozen coaches. The trains on the down line from Cambridge were often late because (it was said) they had to wait for the London train before starting their all-station journey to St. Ives, March and Wisbech. It was not until I had to use them when in 1918 I began attending school in Cambridge, that I was aware of any irregularities. Till then a train journey anywhere was an event indeed. Once or twice I was taken by my mother to Fordham where all my grandparents lived. This was on the Mildenhall single dead-end line from Cambridge where we had to change. On this line they used a 2-4-2 tank engine with a big cab and a high brass-topped chimney. At least this is the one I remember best. After the first journey I would often sneak upstairs and, with eyes shut, work my mother's treadle sewing machine, imagining myself to be driving an engine. I longed to have a picture book of engines, but somehow requests for Christmas or birthday presents in advance of the day were not the done thing in our family, and one did not come my way till I was much older.

As it was, I was sometimes content in those early years with the trains I used to plague my father to draw. Harassed as he must often have been by work, he would seldom refuse if I caught him at his desk or after meals. If I could find a long piece of paper, or a long envelope which could be doubled by slitting the sides, he would begin with an engine outlined in pencil. Then would come the trucks, and with a guard's van at the rear, these had to take up the full length of the paper. But when he had drawn twenty or thirty trucks, it became so much of a habit for me to say that these were too much for one engine to pull, making double heading necessary, that he came to know this in advance and to leave room to fill in the leading engine on demand. Such drawings may have been hurried, but they satisfied me at the time; my father's real hobby was painting flowers and landscapes in oils, which he kept up till his busy life came to an end when he was eighty-six.

In those days you could hear a traction engine on the move a long way off. There were two threshing sets that served the village, but the pattern of Over was mainly in the form of squares or blocks. You could hear an engine clattering in the distance but had no idea which way it would go, and many a time I'd run off only to find it had taken a turning that led to a more distant farmstead. When one could be tracked down, it depended on the good will of the farmer concerned as to whether one would be allowed in to watch it set the drum and strawjack. I came to know this kind of thing and if, when the tackle had entered a yard, the gate was shut with a gruff 'Don't want boys in here', that man went down on my mental black list.

The good farmers were those who tolerated boys like George and me. When a stack was getting down as the sheaves were pitched into the moaning clattering drum with its haze of dust, we might even be allowed up with sticks to kill off escaping mice. These we would stuff into our pockets to take home for the cats, or to plague the girls at school. That was when we were at Over Church School, and more than once I was caned over the hands for placing dead mice in places best calculated to cause a fright.

The greatest joy from threshing days came when I was allowed to sit on the tender of the engine. The drivers of both engines were kindly disposed to me, although I'd stand perhaps for half an hour watching the motion from below before being beckoned up. Once there, I became enthralled by the sight of all those parts, moving in unison with such power, and by the smells which were so much more concentrated from this exalted position.

Seamark's Fowell engine was newer and better kept than Mustill's 'Marshall'. It was dark green and its brasswork shone, but Albert Mustill's was grimy; and even though it went well enough, it was generally accepted that Mr. Seamark's (he came from nearby Willingham) was a grade better in every way. He was quiet and methodical, never flustered even when in the thick of muddy conditions which so often prevailed. But Albert Mustill was always in trouble, it seemed, and people said he spent too much time and money in one of the twelve pubs in the village. Still, there was nothing he would not tackle. If one of the old thatched houses or sheds with their clay-bat walls needed demolishing, Albert Mustill would do the job. He'd stand his engine strategically with heavy chocks behind the rear wheels and would play out the wire rope to encircle the building a foot or two above ground, hooking the end back on to the engine. Then as the rope slowly tightened it began to squeeze in till it cut into the four corners of the building. Timber would begin to crack and the roof would begin to quiver, till with just a cough or two from the engine the whole structure collapsed in a cloud of dust. And then Albert Mustill would give a mischievous grin and say, 'Now that's set somebody else a job.'

Boys were kept well back when such Great Events took place. The downing of a large tree, roots and all, was another sight to be seen from a distance, but mostly it was school that was to blame if any were missed. Albert had few qualms about taking his engine and tackle down to some outlying farm down one of the Fen Droves, deep in ruts and mud. But these farmers had to be more choosy, because ruts and mud would be so much deeper afterwards as to become impassable till drying spring winds came. It was when Albert employed Frank Parish as driver that more chances came my way to squat happily on the coal in the tender and watch. Fifty years later he came to see mine.

Albert was an opportunist, and when in March 1918 a biplane crashed on Fred Kirby's farm, he promptly took the struts for repairing his threshing drum. Other parts were taken by other people, following a rumour that the authorities had no objection, but before nightfall the village Crier was out, announcing severe penalties to all who did not return everything immediately.

The prospect of seeing more railway engines was some compensation for having to go to school in Cambridge when I was nearly twelve. The 8.14 a.m. at Swavesey reached Cambridge, eleven miles or so distant, just in time to clear that immensely long No. 1 station platform for the 9.5 express to London coming in from Ely. At the two shorter inner platforms at the south end could be seen L.N.W.R. trains which ran to and from Bletchley, linking Cambridge with Oxford, and the Great Northern spur to Hitchin and King's Cross. None of the engines on either of these was large and some must have been very old, with scarcely any cab to protect the driving crew. But the green G.N. especially were interesting, and more varied in type, though none of the type I longed to see in real use, with outside cylinders. It was years, in fact, before I saw one on these Cambridgeshire lines, but I'd by now seen them as pictures. Among these was the only outside-cylindered locomotive the Great Eastern had built for donkeys' years, the remarkable Decapod, which for some inexplicable reason had been discarded very soon after making its appearance as a prototype.

Nevertheless, Cambridge station was for me so full of interest that very often I lagged behind the rest of the boys who tramped through the goods yard as a short cut to the High School over Hills Road railway bridge. The railway ran at the far end of the school playing field, and trains could be seen from most of the classrooms facing that way. The largest engines were known as 'Big Sixes' (the 1500 Class) and there were 'Big Fours' as well as 'Little Fours' on the Great Eastern. It was an *event* if a Big Four (Claud Hamilton) headed our school train. But the Big Sixes were reserved for main-line running. Almost invariably one of these would be waiting on a short siding to take over the 9.5 to Liverpool Street.

With very little grace allowed us for the 9 a.m. start of the school day, there was precious little time to linger and watch the change of engines. A quick appraisal of the waiting '1500' was not enough, and when (perhaps once a month) Driver Ablett was in charge it was hard to resist the urge to linger. Of the waiting drivers who might be there when the 9.5 came round, he was by far the friendliest. He'd even told me his name, and with no real hope of reward I began taking him little gifts of fruit and flowers from our garden. We became friends and then one day he beckoned me over as our train drew in. 'The Inspector's just gone back,' he said, 'come up and have a ride when the 9.5 comes in.

I know you won't split on me.' I could have hugged him, but instead crouched out of sight, thrilled to the bone. Five minutes later the train for London drew in, and with its engines uncoupled and away, Mr. Ablett pulled the regulator and away we went up to the points and backed on to the 9.5. It was worth the 'impot' meted out for being too late for prayers with too lame an excuse. The chance never came my way again, but the pride it gave me of being the only boy to whom this honour had been granted took years to wear thin, and it made Mr. Ablett my friend for life, in my own estimation.

During the 1918–22 period when I was at school in Cambridge, the railways were said to be suffering from the effects of the Great War. Cleanliness and time-keeping had slackened, but most of the G.E. passenger locomotives were still in their deep blue livery beneath the grime. Only on the London trains were there corridor coaches, until the latter part of that time, and these were often clerestory roofed. It was commonplace to see a man at dusk on winter evenings on Cambridge station walking along the roof of a train with a flaming torch, lighting up the gas lamps of the compartments below. Sometimes, when it was very cold, tin hot-water foot-warmers would be pushed inside each compartment, causing a scramble among the seven or eight boys who, having collared an 'empty', used both guile and intimidation to keep out strangers who might spoil any fun and games they had planned to begin once the train left the station. We were not prone to vandalism, but window straps were known to come adrift if an older boy tried to enter the preserves of a bunch of younger ones. And if one of our number was caddish or became a culprit for some reason, he would be sentenced to imprisonment for the duration of the journey. For a misdemeanour he would be hoisted into the luggage rack, but for some greater offence would be pushed under the seat and kept there very effectively by several pairs of heels.

Brass soldiers' buttons were fairly easily come by at that time and one of these placed on the line would be flattened to just the correct size for most of the slot machines on Cambridge station: that is, until they improved the design so that the mechanism could make the distinction between brass and bronze. Very occasionally, during my later school days, the train would have corridor coaches. Everything had to be explored and tried out, and once I discovered that the door next to the engine tender was unlocked. With no thought of danger, any discomfort there was in travelling between stations was offset by the thrill of seeing the flashing track beneath my feet as I stood there on the buffers and coupling, so close to the engine that I could hear the fireman shovelling. Trains and engines were part of life during those few years. With the loco sheds just beyond the wall of our departure platform for home, sometimes, especially on hot summer days, the smoke would be almost overpowering; but the smoke from our own homeward-bound train was sweet in more senses than one. Only once did we see an engine in trouble. A G.N. incoming train overran its platform and its front wheels had pushed down the wall of a parcels office. There it stood with chimney touching the station roof; its great bulk at an alarming angle, still hot with escaping steam, brought me once more that feeling of awe.

3

When Steam was Commonplace

Schooldays for me ended prematurely, but at least this was of my own choosing. I wanted to do things, to work and make my way in life. My father's market garden was his first love, though it was his secondary business. Flowers appealed to me too, and as I had passed through the period when I wanted to become an engine driver when I grew up, I decided to go into horticulture at about the time that my parents moved from Over to Oakington, five miles nearer Cambridge. The holding there gave my father the chance he'd been waiting for, of concentrating on fruit, flowers and plants, and for a time I worked there with him. Full of big ideas, but with very little knowledge, it took a three-year spell away from home, working on nurseries in Kent and Yorkshire, to help to clip me down to size, though I was left with ambitions unscathed.

One of the disabilities of youth is the belief that the road of life beckons endlessly on, allowing time and opportunity for every hope and ambition to be fulfilled. At least that was how I saw it, and acted accordingly. Having crystallized my ambition to build up a nursery, rearing and selling outdoor plants rather than going in for my father's type of market gardening which seemed so chancy, I kept it as my fixed goal as I approached manhood. Apart from such sports and recreations as football, tennis and ice skating I tried to squash other interests which might hamper, including my delight in steam engines. Not that I could ever resist the sight of one, whatever else I might be doing, and I had to know in advance of any rail journey I took what kind of engine was heading the train.

Our place at Oakington was only two fields away from the St. Ives–Cambridge line, and when I went to work on a nursery at Tunbridge Wells, trains were even closer. The ex-L.B.S.C. line to Edenbridge skirted the nursery on an embankment which was also level with my bedroom window not far away. I lodged with a railwayman who had often joked about the job of tapping carriage wheels with a long-handled hammer. But the locomotives on that line were not specially interesting, mostly 0-4-2 tanks of considerable age which both pushed and pulled their coaches. From the former S.E.C.R. station near the centre of Tunbridge Wells larger engines, though mostly 4-4-0's, took faster trains to London Bridge or Cannon Street stations. But still the sight of an outside-cylindered loco eluded me until early in 1925 I applied for a job on a Yorkshire nursery. As I had never been there the north country appealed to me, and not least because I'd been told that that was where one had to go to see a great variety of big engines. With self-confidence far outstripping actual experience, I gained additional bolstering from a seat on the 10 a.m. from King's Cross, hauled by an A. 1 Pacific. The name of this great and resplendent locomotive I do not remember, being so overawed at the sight of it. And at York I had such a feast of engines that I became all the more determined to gain such further experience as I needed in Yorkshire.

I was accepted for the job, for which in truth I was too young and inexperienced, and by missing a connection on the return journey I spent six hours on York station next evening; there were too many railway engines to be seen for me to think of trying to sleep in the waiting-room.

Nostalgia takes over only when pleasurable sights and experiences are no longer available. And when one gets over the hump of life's span and the past recedes, the memory strives to bring it back again. My memory is nothing if not vivid, but it lacks breadth so far as engines are concerned. That period, before the stark realization came that the steam age would inevitably pass away, can only be covered here as an exercise in nostalgic memories. I can but deeply regret lost opportunities for seeing steam in action during its heyday. Yet there is more than the one great consolation of having a collection of engines within a stone's throw of my armchair, because what I did see of steam in younger days left an indelible impression on my mind. I have no valid explanation for my preference for locomotives with outside cylinders, and have never bothered to define it. For all I know it might be psychological, and quite likely someone with a Freudian mental approach could explain it in reduced or basic human terms. Whatever the reason, I can take in the sight of an inside-cylindered engine in a matter of seconds whereas it is an effort to take my eyes away from one with both outside valve gear and cylinders.

In Yorkshire I saw these and many others. The amalgamation of 1923 had led to increased numbers and greater variety on lines that had previously used their own motive power exclusively: the Great Northern, North Eastern, Lancashire & Yorkshire and others. When I returned to Cambridgeshire in 1926 to work up a plant production side to my father's business, there was disappointment in finding the same old G.E. stock running on the railway at Oakington. The passenger trains were nearly always headed by black 'Claud Hamiltons', whose sharp, whistly beat following the 'plock' of the snifter valves was so distinctive that one scarcely troubled to glance over the fields to see them pass on their way. The frequent coal trains were still hauled by the uninteresting J. 16's and 17's. Though some of the latter now had larger boilers, they were still plain 0-6-0's with a top-heavy appearance which would have been obviated and greatly improved if only they had added a pair of leading wheels and had fitted outside cylinders. To me, it was as if the G.E. was deliberately lagging behind in this respect.

Within weeks of returning home I found myself offering to become a fireman or guard at a recruiting depot near Cambridge station. The call had gone out for volunteers to break the General Strike. As there were no flowers to be cut for distant markets, because there were no train services, my father raised no objection. In any case I was nineteen and the idea appealed to me greatly. But University students had got there first, for though my name was recorded I was never called.

Some months later, back home and at work again, I heard a strange sound coming from the direction of the railway. It was from the down side of Oakington station, and though obviously from a fairly slow-moving coal train (the numbers of which had increased since the strike) the rhythm intrigued me so much that I downed my fork and ran across the intervening meadows. It was not so much the exhaust beat that excited me as the 'clinker-bolcher-clinker-bolcher' sound, and then at last I saw the K. locomotive come into the open on the up side of the station. At last! Here was an engine with outside motion in service on this line, and as I leaned against the railway fence, I was completely enthralled by the probability that from now on others would follow. And so it proved, and though belated, this innovation on the G.E. section of the L.N.E.R. came as a decided boost to my interest. Since it was an innovation, it was some time before I knew its official type, and the name 'clinker-bolcher' became thereafter part of my vocabulary.

4
Engines Elusive

A family decision in 1927 to switch from horse to motor transport had some reaction on my fringe association with steam. Although I cycled greater distances than we ever travelled by horse and buggy, the acquisition of a van was a novelty, even if it was only a solid-tyred Trojan, and though it cost a mere £135 new it had to be paid for by instalments. Money was tight, and with total sales in 1926 of only £650 from the six-acre holding, my plea for a van with which to increase sales must have caused my father much concern, with profits barely above subsistence level. The van helped a little, but as its only driver I was the one to go to markets and stage exhibits at distant shows; for a time, then, roads became more interesting than railways. It was a kind of compromise, for though I felt strong urges to travel I was by now too deeply committed to working up a hardy plant business, and at that time I was the only one of three brothers interested in horticulture. Eric was in Australia and George working in London.

I made time, of course, to linger near main-line level crossings, embankments and bridges. But this practice had its snags. I'd feel twinges of conscience while hopefully waiting for a train to appear, knowing that I really ought not to waste time, since time was money. And then if and when a train appeared in the distance I'd forget that aspect in the mounting thrill and excitement. If it were an express that came roaring past I'd feel a lump in my throat, and often have to close my eyes because the sight was almost too much to bear. It was not only the noble magnificence and sheer power of the locomotive, but the fact that it was hurtling into distant places without me. It was the G.N. main line through Huntingdonshire that drew me most, and despite the unsettling effect of the sight of express trains, I could never resist any opportunity that came of going back for more.

I drove that Trojan van 30,000 miles over the next three years, and though some progress was made on the business side, it became clearer to my father than to me that his interest in flower growing and indoor plants would come into conflict with mine in hardy plants and alpines. This conflict was sharpened when in 1929 my younger brother George decided that he too wanted to become a nurseryman. His preference was in line with my father's, and next year came the split. What it cost my father in terms of courage and forbearance I shall never know, but at the age of sixty he took over a derelict glasshouse nursery at Mildenhall, and when he'd spent a month there alone making the house habitable he moved in with my mother and brother. I was left behind with my opportunity when I was barely twenty-four; the challenge it carried was no burden even if my working capital was only £50 and my rent £75, and when I married in April 1931, the future appeared very rosy indeed to me.

Such autobiographical details may appear to have no direct connection with the title of my narrative, but they are part of the background to it. They are helpful too, for the coach driver mentioned at the beginning may not be the only one to come to erroneous conclusions on a first visit to Bressingham. Times were hard, and for many years before as well as for several years after 1931, pinching, scraping, saving and making do were a real necessity, and some of the habits this stringent period imposed I have never been able to drop.

The depression of the early 1930's was not a propitious period in which to work up a business from scratch with virtually no capital. Luckily I'd inherited my father's zest for work, and this, combined with an urge to specialize, led to my decision in 1934 to produce plants solely for the wholesale trade. By 1939, my ambition to have as good a hardy-plant nursery as there was in the country was largely achieved, with further objectives cropping up until the war came to blanket them down. By then I'd acquired a derelict farm, twenty miles away. The idea had been to compete with the Dutch nurserymen, since it was similar soil to theirs but with poor drainage. But with no sale for the thirty acres of plants now under cultivation at Oakington, and six acres on trial in Burwell Fen, the necessary switch to food production put me right back to hard times again.

This part of the story has been told before,[1] and since it has

[1] *The Farm in the Fen*, Faber & Faber, 1944

nothing to do with my title I must revert to incidents during the thirties that kept me, if only incidentally, in contact with steam.

It was in 1934 that I felt able to replace a hand pump for watering nursery plants with a petrol engine pumping into a water tower to give the necessary pressure. Having bought at second hand a dismantled steel girder tower, thirty feet from base to the tank on top, the problem arose as to whether I should re-erect it on its new site or assemble it lying on the ground and pull it up afterwards. Either way presented difficulties, but when Harold Papworth, a friend and neighbour who owned steam threshing and ploughing sets, offered to pull up the tower with one of his engines, my decision was quickly made. With a concrete pad ready, the tower was raised by jacks as high as we could get it before the engine arrived. So were some sheerlegs, which Harold explained must be used to gain leverage. Such operations as this had been quite outside my experience, and with the sheerlegs anchored beyond the pad the tower was raised higher, but only a little. More bolstering was needed because the angle was still so great that the legs of the tower were digging in, and in danger of buckling against the concrete of the pad. It was hours before the angle narrowed sufficiently between the top of the tower to which the engine's winch rope was fixed and the snatch block pulleys on the sheerlegs through which the rope ran. When at last the pull was correct the whole operation nearly ended in catastrophe. A little too much throttle on the engine as the tower reached the vertical caused it to totter drunkenly on its spidery legs, and it all but fell over the opposite way before the base bolts could be secured.

It was two years before another of Papworth Bros. engines had occasion to come in at the nursery gate. By then, the water tower as an installation for irrigation had become inadequate. I'd invested in spray lines, another second-hand bargain, and for these much greater pressure was needed. So was a new well that would provide sufficient water for a much larger pumping unit. Hole digging, which came to be almost a recreation for me if I found an excuse to dig one, had revealed that water was fairly abundant at ten to twelve feet down. The well I'd dug for the water tower had provided some practice, but the new one had to be much larger and deeper than this, and sited nearer the centre of the nursery. Having reached water in the gravel at about ten feet, I inserted an old tumbril cart wheel and began laying bricks on this in the usual manner so that they interlocked. As I dug down, wearing thigh boots, filling buckets to be hauled up by hand from above, the brick wall bedded down under its own increasing weight as more bricks were added. The water oozed in quickly from all sides, so quickly in fact that nothing we had could cope with it and thus enable me to deepen the well sufficiently. With an irrigation system needing 5,000 gallons an hour, it had to be deep enough never to allow air into the pump suction valve.

At thirteen or so feet I was beaten. I looked up to see Harold Papworth smiling down at me. 'Looks to me as if you'll have to come up if you don't want to stay down there for good,' he remarked, as another chunk caved in. I was beginning to think that way too, and asked him if he knew where I could hire a portable pump, because I must have it at least another two feet deeper. He said he did not know, but added that one of his ploughing sets was heading this way, and if I'd like to try it, the suction pipe on one of them might help.

Up the ladder and into the sunshine again, and we did not have to wait long before the unmistakable ringing clatter made by Fowler ploughing engines in transit was heard. 'We'll try one first,' Harold advised, adding that both engines would have made more sure of it, had there been room for them. This one began to assume larger proportions as I descended the ladder again, with its suction pipe dangling down into the well, not long enough to enable the engine to stand a really safe distance away. But the head of water fell perceptibly as soon as the driver turned on steam, allowing it to run out of the engine's tank away from the well. As soon as I could, I began again to scoop up sand and gravel frantically, with two men hauling the buckets up on ropes over a pulley. The deeper I dug the faster the water ran in, and with that 20-ton engine looming high above my head, and with no apparent reason why it should not press the soft unshored sides in under its own weight and trap me, my feelings went beyond awe this time. But having taken out another foot or more, and having laid more rows of bricks to what I gauged would be the rest level of water, I asked for the ladder and climbed up much more quickly than I'd gone down. It had been one of those situations when determination becomes mixed with face-saving to such an extent that both discretion and fear are forgotten. It was not until I was back on top, automatically filling my pipe with tobacco, that I realized how foolhardy I'd been.

There were no risks attached to the next operation when steam power was called for. Expansion of nursery acreage had been fairly rapid. A nine-acre meadow adjoining the original six acres had been bought on mortgage a year or two before I began on my own. By 1934 this too had all been taken in and I rented another six acres. But by 1937 I was looking for more land, and as a new airfield was gobbling up so much of the district, I had to seek farther afield. An orchard in the neighbouring parish of Cottenham was to let, but it was played out. Provided that I had the trees pulled up, the owner said, I could have it the first year rent free. Although I knew full well that the cost of this must exceed the rental value of open land, I had no other choice. Likewise, there was no other choice but to employ Papworth Bros. ploughing engines to pull up the trees, roots and all.

With each engine at the end of a row, and helpers with chains to wrap round each trunk, one after the other those old plum and apple trees came out with the pull of the engines as if they had no roots at all. It seemed so effortless, but anyone who has dug out a tree with a spade and an axe to cut through the roots would appreciate the difference when steam provides the power at the end of a wire rope. It took less than three days to pull out nine acres of trees, but it took more than three weeks for ten to twelve men to clear them away.

I'd become land hungry. Already I'd looked around for a larger holding, a farm that would give room for expansion as well as for practising straight farming. But though ordinary farm land could be had for ten to twenty pounds an acre at that time, I could not raise the price for the type needed for nursery work. So it was that in 1938 I fell for the farm in Burwell Fen. Following a prosperous season I was just able to put my hand on the £1,600 for this remote two hundred-acre farm, nearly a mile from a hard road, and much of it waterlogged and derelict.

By then, though becoming aware of my weakness for seeing the end of a project that appealed without fully counting or contemplating the means required to reach that end, I little realized just what I'd let myself in for. The onset of war may have knocked the bottom out of my nursery business, but I sometimes reflected that but for the wartime needs of increased food production I might well have come a cropper in farming. During the latter half of 1939 and the early part of 1940, I had little but hope to work with. I was on my beam ends for capital, and only improved drainage and access to the farm formed grounds for hope, which had to be backed by whatever fight I could put up.

The grim winter of 1940 froze the soft peaty fen 'drove' to the farm solid enough for access, but stopped all land work. In my near desperation the idea of laying a railway track occurred, partly to make a job and partly as a means of better access when the drove reverted to mud. Many bog oaks had been encountered when ploughing the year before, and these I thought could be cut into sleepers. For a few pounds I bought a saw bench fitted with an Austin 7 engine, but I soon learned that bog oak will blunt a saw more quickly than any other kind of wood. The under-powered circular blade generated such heat from friction as to produce smoke from the sawdust.

The output of sleepers for a two-foot-gauge railway was as pitiful and as small as my hopes of being able to acquire track. The price of such few lengths of rail as I located for sale was frightening, and as the long winter ended, so ended my hopes for what had begun as a bright idea. I had scarcely gone far enough to think about the motive power for my railway. Steam would of course have been my preference, but even my joyful hope of seeing steam tackle threshing out my crops was dashed. Other farmers in these peaty fens carried their sheaves to the nearest hard road for stacking at harvest time, and after my first season down there the reason for this became plain enough. These droves might carry the weight of a traction engine in dry summer weather where the drainage was good, but in a wet autumn or winter, no engine owner could be expected to tackle Burwell Fen.

5

Bella Comes to Bressingham

With more foresight I might have perceived, by 1941, that I could have profited far more by sticking to the Oakington nursery with its potential for intensive cultivation, than by carrying on with the Burwell Farm, twenty miles away by road. But as soon as the drainage had been improved in that Fen, so long forlorn, I saw myself as a reclaimer of derelict land. By sticking out my neck, I accepted the offer of another three hundred acres of the wild Adventurers Fen which was adjoining, and for the next two years put all the concentration I could muster into the task of making the waste grow food crops. *The Farm in the Fen* recorded that story years ago as my first book, but it did not tell how subsequently so much of my labours proved vain. That was because it was decided to allow Adventurers Fen to revert to wilderness as soon as possible after the war, since ownership was vested in the National Trust. It was more than I could stomach, and I felt so bitter about it at the time that I gave up my tenancy of it and put my own farm up for sale. The Oakington holding of thirty-six acres was much more profitable during those years than the five hundred and fifty acres in Burwell Fen. The urge to remain both farmer and nurseryman had, however, become fixed, and a move became imperative. By selling out all my interests in Cambridgeshire, I was able to raise just enough capital to buy Bressingham Hall Farm in Norfolk early in 1946.

Of the two hundred and twenty acres at Bressingham, nearly a third was unproductive swampy grazing, or marshy, reed-infested scrub, and tangled woods of alder, willow and thorn. Still feeling bitter about the outcome of my reclamation efforts in Burwell Fen there was the consolation that here any land I could reclaim would remain mine. Then there were also some sound upland fields on the sloping edge of this upper Waveney valley. These should offer scope, I thought, both for straight farming and for the restoration of the nursery business, so shrunken by the war. During the bitter winter of 1946/7, the children had to be taken to and brought home from school by car. It was only two and a half miles to Diss, but the roads were icy and cycling was far too dangerous. One afternoon in February, when the biting east wind was bringing down more powdery snow from a leaden sky, I drove them back from school and turned down the farm roadway instead of the drive to the house. I'd said I had something to show them, and had played a guessing game till they gave up, just before we reached the farmyard. 'Is that what you've been teasing us about?' asked eleven-year-old Bridget rather disdainfully. 'What's it here for?' asked Robert, then eight; but Adrian, a year younger, asked if it would go. I said of course it would, feeling a little hurt that they showed no excitement at the sight of a traction engine. Perhaps the time and temperature were to blame but I'd hoped to play up to an expected climax when I told them I'd bought it. Now the telling had to come regardless of anticlimax, and still they were not greatly impressed, even when I said its first job would be to saw up firewood. They were obviously keener to go indoors to the warmth firewood provided than to inspect the engine standing there cold and lifeless after being unloaded an hour before. And I had to let it go at that, hoping for a better response at a more favourable moment.

There was no doubt in my mind, that this acquisition had come at a very propitious time. Piles of firewood needed to be sawn into logs. Coal was scarce and I had ideas of meeting a need for a cash return, since for months a lot of money had been going out and none was coming in. I had other ideas, too, of ways in which this purchase would prove to be a useful one, even if the simple truth was that I just wanted to own a steam engine regardless. It was a single-cylinder Burrell, built in 1912 by the well-known Thetford firm, and it weighed nine and a half tons. I'd been offered it by a Mr. Pymar of Mellis, a few miles away, who had explained that the 'War Ag.' had acquired it as part of a threshing set they had brought into service. They restored the engine, but it became redundant almost immediately when a new tractor arrived. Though the price of £50 was well above the market value as scrap, it was in good working condition, and a safer buy than an engine with no guarantee. Lack of knowledge, coupled with my long-standing respectful awe of engines, made this an important factor.

When the children had gone indoors, I walked around it and

stood at the controls once again. My farm foreman, George Long, came and stood beside it, gazing admiringly. 'My 'ole dad were an engine driver,' he remarked. 'I reckon I could remember enough of what he told me to drive her.' That was all very well, I thought, but I wanted to drive her myself, and though prepared to share with George to some extent, I thought it best first to seek the advice of Tim Stevens, a local engine owner recently retired, as to how an engine should be treated. At least George approved of my purchase which was some comfort, I thought, as we tied down her canvas tilt for the night.

That evening Tim Stevens gave me the initial advice by telephone, saying he would come along in the morning. By the time he arrived the boiler, now filled with water, was warming up from the fire I'd lit earlier; and under Tim's guidance, the oiling and greasing procedure was carefully followed. Tim certainly knew his stuff, if the long string of do's and don't's was anything to go by, which I dared not admit I found a little bewildering. He was regarded as the best local expert on steam. 'Sixty years I've had of engines', he said proudly, 'and the first time I drove one, my dad had to stand me on a gurt lump of coal so as I could reach.' When I told him that the first job was to be on a saw bench, he said he knew something about that too, and had spent his later years hauling timber with a steam tractor. With pressure at 60 lb. the little man put her into gear, and with a shrill blast on the whistle from me, we moved bumpily over the frosty ground down to the spinney where the saw bench stood. Our one tractor, a David Brown, had not been man enough to drive the heavy old saw bench I'd recently bought, but steam would make easy work of it. It was exciting to say the least and I was as proud as Punch to become the owner of a real live engine at last. If it earned its keep I'd ask no more. If it didn't, I'd still never want to part with it anyway.

Tim set the Burrell in line for the saw bench. Having made sure the driving belt was right for line and tension, from the engine fly wheel to the saw bench pulley, he said he had to go and would return next day. 'You'll be all right,' he assured us, after repeating some of his earlier advice and adding more warnings about the importance of the water level in the boiler.

A small heap of logs for fuel lay nearby, already sawn. Two more men were called in to help George Long feed the saw bench, and with a hundred and twenty pounds of steam the blade began to whirr and whine. Most of the noise came from that circular saw, for the engine was quiet enough. The cranks ran smoothly and silently as the piston rod drove in and out from the cylinder and its only sound came from the gentle coughing beat of the exhaust from the tall copper-topped chimney. It sounded like a soft 'bella-bella' rhythm and from that first day, 'Bella' became her name.

This was marvellous, I thought, standing up there at the controls. At least to begin with, till I noticed how George and his helpers were having to push the cordwood on to the saw blade. A few logs came off, to add to the little heap of engine fuel, but at the end of an hour there was scarcely any more on the pile than there was to begin with. The engine had in fact consumed almost as much as it had sawn. True, I'd taken rather a delight in stoking, and though the wood was damp the pressure gauge was nearly up to the red mark of 150 lb. at which the safety valves blew off. It was puzzling and worrying. We'd blamed the lack of output, when using the tractor on the saw bench, to lack of power and I'd been banking so much on the Burrell doing a more efficient job. There was enough cordwood to go at, piled up ready, to bring in £50 or more, sold as logs with fuel so scarce just then.

We switched to thinner boughs, for it seemed that the thicker pieces of elm and oak were too tough. But though output improved a little, something was obviously wrong. Just as we decided to give up for the day, Tim Stevens came along, saying that as he was passing by, he'd just see how it was going. When we had told him the symptoms, he went to the bench and fingered the friction-hot blade. 'Here's your trouble,' he announced, 'the blade needs setting as well as sharpening,' and explained that the teeth were in need of a special tool which made for a wider cut and so would reduce friction even if it made more sawdust. Thinking of bog-oak sleepers at Burwell I saw how, once again, my over-eagerness to plunge into jobs without knowledge and experience had led to failure. Within hours, I learned later, all the staff and most of the village were making a joke of how Alan Bloom's engine burned up more firewood than it cut. By then I could begin to see the funny side of it too, because with Tim's help the blade was set and sharpened and the heap of sawn wood was growing fast. Yet despite the fuel shortage, an advertisement for firewood brought no replies worth bothering about and we had to switch to sawing up 'two by fours' and fence posts for use on the farm.

The winter dragged on until mid-March. Then came a gale which brought down nearly fifty trees and odd bits of roofing. The heavy rain coming as it did with frost-locked ground and snow-filled ditches flooded over a hundred acres of the farm. From then till mid-May we struggled to free waterlogged land so as to plant up the new nursery. But the next instalment from that perverse jade, the weather, was drought and for weeks we struggled to keep alive what had been planted. It was not until July that Bella the steam engine could be used again. By then the peaty fen in the valley bottom, which had been under a foot or two of water in March, had dried out enough, we reckoned, to take her weight. Down there were over twenty acres of derelict land marked for reclamation. Most of the bushes had been cut back during the winter frost, but stumps remained, along with a variety of smaller trees of willow and alder which must be cleared. Bella's job was to pull them out so that the land could be ploughed for the first time for over a century, according to local reckoning.

There had been no time when the frost broke to give Bella a boiler cleansing, but it had to be done before she was put to work again. George and I decided to drive her a hundred yards to the nearest pond, since there was no piped supply. With steam up, it dawned upon us that in spite of having spent a month with her as a stationary engine driving the saw bench, neither George nor I had the least experience of driving the engine as a self-propelled vehicle. And with two gateways, a short slope, and a narrow ditch-bordered lane to negotiate, we decided that if two were on the footplate it might be safer. George took the throttle, but soon showed that what his dad had told him about driving was too far back to remember. I took the steering wheel, but found that this was scarcely the place in which to find out how many turns it took to drag the chain-held front axle round for a given change of direction. Within twenty yards a gatepost went down like a ninepin. I'd shouted to George to go slow, because of a downward slope, and thought he had shut off steam, but the engine still went forward, as the slope continued. We were nearly across the lane and a deep ditch yawned ahead. 'Stop her,' I shouted again, frantically twiddling the steering wheel to head Bella into the sharp right and then left hand turn to get her over a culvert and into the opposite field. George was obviously trying, and we both looked about for a brake that wasn't there. 'Can't stop her no how,' warned George, and though she began losing momentum, it was too late. Down went the front wheels into the soft but dry bottom of the ditch, with the chimney pitching forward at an undignified angle, and George and I jumping clear from the tender.

'How the 'ell did that 'appen?' George scratched his head with a rueful grin. Though we both felt it was the other's fault, no accusations were made, for the real truth was that neither of us knew enough about engines yet. What we did realize was that with the front down the firebox crown was in danger for lack of water, and though the fire was low we soon had it shovelled out entirely. Tim Stevens to the rescue again, I thought, and when he came he first had to explain how we'd erred, as is the wont of experts. 'The reversing lever's your brake on this engine,' he said, 'but you need just a bit of steam on the throttle to hold you. No steam no back pressure, and being in gear would have stopped you dead if you'd just pulled this lever back.' There was more besides, until the task of recovering Bella ended four hours later with the aid of jacks and timber baulks.

It was with humble thanks that I accepted Tim's offer to drive her down the Causeway to the Fen, once the boiler wash-out had been undertaken. This mishap had shaken my confidence in the art of mastering an engine I had so badly wished to acquire. The Causeway was straight enough but narrow, bordered with trees as well as ditches. But towards the bottom end an old brick bridge over an extra-wide ditch had to be crossed. It was only nine inches thick over the arch, and for safety we'd placed all the strong planks and poles we could muster, so as to spread the weight. Even so, its width would give only six inches' clearance for Bella's six-foot iron wheels. George and I went ahead, when with steam up Tim backed her out of the meadow and headed down the Causeway. She was a pretty sight, with the white wood smoke and steam mingling with the greenery of the spreading trees above. Wizened, tiny Tim could scarcely be seen back there at the controls, but when twenty yards from the bridge he stopped and climbed down. 'She should make it,' he said cheerfully as he surveyed the hump of timber covering the bridge. 'She'll fall soft-like anyway, if she does break through.' Such sang-froid struck me as being rather too flippant. The drop from the bridge down to the water was ten feet or so, and below that was deep black mud. Of course she would fall soft-like, but what were the chances, I wondered, of raising her again, to say nothing of Tim's welfare. But the chance had to be taken. Slowly mounting the timber, Tim gently drove her over. His tense expression quite contradicted his tone of a few minutes before. A couple of bricks squeezed out and splashed into the water below, and then with a flourish and a triumphant blast on the whistle Tim opened the throttle, the danger past.

A couple of hundred yards farther on was the Waveney river, not more than a wide, steeply banked ditch at this point, three miles from its source. There I'd dug a channel leading to a firm piece of ground beneath the trees, to provide water for Bella's tanks. It was in a thicket of trees and evergreens and from there I'd planned to take her out daily for the next few weeks to clear this wild, tangled twenty acres of its scrub and useless timber growth. It proved to be satisfying work, for with scotches behind Bella's rear wheels, the wire rope was paid out to helpers with chains.

The technique was simple. Once a chain was looped round a bush or stump, a raised hand was the signal to pull. A touch on the engine throttle began to wind in the cable on its drum, fixed as a loose pulley to the near side hind wheel. The chain tightened and when all was taut, Bella would give only a little cough before out they came, roots and all, from the damp black soil. Mine was by far the pleasantest task, and though George was relief driver, I reckoned that at last I was learning how to handle and care for an engine.

Those on the other end of the hawser did not complain, except to swear at the tormenting horse flies which settled on bare flesh so gently that only their sting was felt. It was sweaty work, and soon great fires were consuming the brush and roots, leaving an ever-growing space of open land. It took one man most of his time to collect pieces of wood for Bella to consume, yet to be self-supporting for fuel seemed right.

The summer was proving hot and dry, just right for such work, yet there was one hazard which after a few days made itself known. It was Bella's nine-ton weight that caused it, and very soon I found it unwise to use the same track repeatedly. The first few acres were so thickly choked with sallow bushes that once they were yanked out the soft black earth held no surface cover. In a few open patches, reeds and sedge grew and their roots held the soil. But water was not far down and when an alder or willow was pulled out, it brought a huge ball of soil clinging to the roots. Into the hole thus made, water quickly seeped despite the drought, and the danger of Bella's becoming bogged caused us increasing concern. The farther out from her station the more tricky it became to keep her on firm ground. The time came when we had to cross an old overgrown ditch, and it took a score of thick alder poles laid longways to the ditch to make it safe. And as these were steadily squeezed down into the quaking soggy peat, others had to be added.

Yet we were making progress, and after so many trials and tribulations of the past few months meted out by the weather, we were making good use of it now, even if the nursery was still suffering from drought. Bella gave no trouble mechanically. It took much less than two hours to raise steam from cold, using the dry, quick-burning wood to be had for the picking. With three men piling the brush on to fires, three more on the grubbing, we sometimes cleared more than a half acre of ground in a day's work.

It had taken five weeks to clear these twenty acres. An additional task had been to dig over each day the area where a fire had burned the day before. Peat is combustible, as I knew well enough from similar conditions at Burwell. Once a surface fire dies out, the peat beneath it will ignite and burn slowly and insidiously downwards to water level. The area was not entirely of peat, and one or two low sandbanks had become lifelines for access to the work, since they were stable and would carry Bella's weight. At the end of five weeks only a little patch of bushes remained in the middle of twelve acres which generations before had been a field. This patch had been left till last because it was out of reach of our sixty-yard hawser, but now it had to be tackled. On several occasions one of Bella's big rear wheels had gone down into the black ooze. It was useless to put on steam and try to make her drive out in gear, for she would have churned and gone in more deeply than ever. The remedy was simple. By hitching the rope to a convenient tree or extra-large clump of bushes, in line with the pull, we enabled her to pull herself out with no trouble at all as we wound the rope in again, with the wheels free of the gears. But now, as we had worked inwards from the outside, this remaining patch of scrub offered little or nothing for good anchorage and the approach was very soft.

I was determined that Bella should be used to the last, so bushes were laid over the softest spots and the most careful approach was made. Shunting was hazardous and in order to reduce movement, water for Bella was carried in buckets from the nearest hole left by the removal of a larger tree. With only a day's work left, it fell to George to do the morning shift. When I took over, he admitted that Bella had gone down and for lack of good anchorage it had taken an hour to get her out, so I'd better watch out because in the process he'd pulled out the last stump capable of acting as anchor. I'd keep on top, I thought, as I took over for the afternoon, the last few hours of the operation. But as I walked through and round the patch I was not so sure, because I could find nothing that offered the security that might be needed. Shunting gingerly, so as to get in line with what remained to be pulled, we managed to get most of the scrub out before having to take up the final stance. I backed on to what I thought was a firmly rooted clump of sedge, which should hold. But when it was too late I realized that the sedge had been thrown into a hole and only appeared to be growing there. As the wheel began to sink in I rashly opened the throttle in hopes of keeping on the move. But it was no use, Bella heeled over and up came her chimney. All I could do was jump off the tender lest she fall on to her side. 'Any stumps left?' I called to my helpers. But now they were as helpless as I. We examined the few remaining and decided that the only chance lay in hitching to one that was lying not quite in line ahead. We made the cable fast, and holding on very precariously with such a list, I gently pulled the throttle. The rope tightened. Bella began to move and rise with ooze dripping from her wheels, and then suddenly the stump came up and back the engine settled, more deeply than before.

Having told my helpers to keep her fire going, I walked back to the house, downheartedly. An hour later I came back with a new length of rope for which I'd paid £10, as the only means of getting Bella out. Thinking of saving costs I reflected sadly that had I known, I'd have bought an extra length a week or two ago, for this £10 might well have been saved by using it to keep on firmer ground when clearing those last few difficult areas in the middle of the field.

George saw me with the new rope and joined in with some sly but caustic remarks as to my efficiency. Though eager to help he made the most of the occasion, till I took umbrage and asked him if he wished to take over; finally the double length of rope was fixed to a sturdy birch, a hundred yards away. 'Not me,' he retorted. 'Them as gets 'em in should be the ones to get 'em out. I will if you tell me I must, but not unless.'

It seemed impossible that Bella could pull herself out, even if I could keep a foothold at the controls. But though at first I thought she must surely fall on her side, and was prepared to jump clear, she took herself in hand. Once on the level, I let her wind herself all the way to the firm ground beside the quivering anchor tree.

Though scared at the time, once Bella came out of the slough I was glad enough to have been the one to get her out, and it was a step towards becoming her master. It served to deepen my affection for her, but during the next six months I sometimes had to beat around for an excuse to put her in steam. By the time the spring rush of nursery work for 1948 was over, there was no need to find an excuse; this time we had to saw a pile of planks from a dozen tall spruce trees which the gale of March 1947 had laid low in the Spinney. At the end of the few days spent on sawing wood, we decided Bella was due for another boiler-cleanse. With just enough steam to reach the new pond nearby, George and I agreed to open the firebox mudhole doors after tea. When I came back at six, George was already there, using a spanner to release the water. Casting an eye on the pressure gauge, I noticed the needle still showed 5 lb. p.s.i. and said I thought we ought to wait a bit.

'Five pounds is nothing to worry about,' replied George confidently, adding that the hotter the water was the more efficiently would it bring out some mud as well. This I knew was reckoned to be true, and did not argue the point as George took off the nut and bridge. He stood up and gave the plate covering the little hole from inside the firebox shell a poke with his spanner. Out came steam with such a rush that George was completely hidden. I heard him give a yell of pain, and skirting the pall of steam, caught sight of him running away, half crouching as if the devil was on his tail. When I caught up with him he was clasping an oak tree trunk, his face twisted with pain. He was in bed for a fortnight after that, with his legs and groin scalded raw. If it was a bitterly painful lesson for George, it served to remind me again that even at only a little pressure live steam is a force always to be reckoned with, and that it's far better to be safe than sorry.

That was the last steam ever to come out of Bella, though I wasn't to know this at the time.

6

Bella's End and Hopes for Her Successor

Any resilience and resolve I possessed when I so hopefully moved to Bressingham in the autumn of 1946 had sadly sagged within a year. In truth, it was not solely the buffetings dealt by that terrible winter of 1946/7, or the heartaches that came with the drought that followed. My marriage had been shattered a few years before, and I was domestically unsettled. Thoughts of emigrating sprouted like seed and I even convinced myself that the three children who had been my sole responsibility since 1940 would stand a better chance in Canada, a land which had always appealed to me. The urge to go had become so strong during that unsettled period of my life that obstacles merely served to increase my determination. We arrived in Vancouver, British Columbia, in September 1948, but it turned out to be an expensive and foolhardy venture and we returned to England a little under two years later. Fortunately I had Bressingham to return to for, despite its concentrated heartaches, I hadn't been able to bring myself to cut all ties by selling out: I had felt that one day I would want to return. Before that day came, I had been warned by my bank manager that it was nearly on the rocks under the manager I had left in charge, but this was largely my own fault for leaving the place.

The accident to George had been a reminder that although Bella the engine was only one of the things at Bressingham I'd miss, she particularly needed to be treated with respect as well as care; but what I wanted to know first, when I returned from Canada, was just how badly the business had fared. I had to size up how great a task it would be to restore it. Within an hour of my setting foot on Bressingham soil once more, I knew that the manager I'd left in charge would have to go. At heart I knew he was not entirely to blame, if I looked at things in one light. If he had failed in his duty, so had I, by leaving Bressingham to someone not strong enough to resist the temptations my absence presented. But I did blame him when during my first prowlings, field by field, I walked through the spinney. The saw bench was still there, with its blade rusted, but Bella was not. Then not far away I saw a heap of old sheet metal, black and oily, and the sight of it made me wince. Those were Bella's boiler casings, which with her registration plate among them could only mean that she had gone. A surge of fury was almost overwhelming.

I searched for George and found him in the barn. 'What's happened to Bella?' I demanded. George had written to me a few weeks before and confirmed from the staff angle what my bank manager had told me of the financial state. George was the only one on the staff to whom I'd confided that I was coming back, because I wanted to surprise the culprits.

'I dussen't tell you afore,' he admitted, 'I didn't want you to get too upset for a start.' I asked him to come out with it. 'Went for scrap,' he admitted. 'I told 'em you'd be riled when you got to know. And I told 'em there was nothing wrong with it. They said it were no use keepin' it when some diddikoy sort of blokes came about a month agoo and offered twenty quid for scrap. So they set to there and then, and cut her up with gas bottles and smashed her to bits with sledge 'ammers. One of 'em bust his finger through to the bone and asked me to bind it up, but I said no and it serves you b—— well right. I were that mad to see what they did to the ole gal. And I doubt if they've paid for it yet neither, 'cos the manager asked me only last week if I knew who they were. It seems he din't draw the cash, 'cos they said they'd pay when they came for the last load. Well, they took two loads of heavy iron and you've seen what's left for the last load they never took. I kept away, so I never knew who they were.'

My fury remained and I longed for vengeance. George and I could see well enough how those roving scrap dealers had never intended to come for the last load, since it was sheet iron and odds and ends too light to be of value. In relation to other losses caused by negligence, this was of small account, but to me it assumed a much greater importance. The manager had no right, under his terms of employment with the Limited Company I'd formed before leaving for Canada, to sell other than produce without reference to me, since I was still the head of the firm, and its remaining assets were virtually all mine.

I could have sued him at law on more counts than one, but it was this one, the senseless disposal of Bella, that hurt a very tender spot. And over this one loss I made him pay £40, deciding it was best to try to forget the rest, and then get stuck back into a venture I should never have forsaken. That £40 was only a token and no consolation at all for the loss of Bella. I tried to track the scrap dealers down, but scarcely a clue existed. The police also tried to help, but their efforts as well as mine ended in failure.

First things had to come first. I knew that it might well be years before I could set out to buy another engine as replacement for Bella. There were very few left now, it seemed, which had not gone to the scrap merchants who since the war had nosed out and destroyed so many. But because I had neither the money with which to indulge my whim, nor the time to search for one, much less to care for one, I had to squash my feelings for steam till better times came, if ever they did, and if any engine could be had by then.

With the place beginning to respond to the stern efforts I had to make, immersion in work began to have some compensations even if the children saw too little of me. Hard times, both in Canada and back home, had hit them too, but they made no complaint except to show occasional signs of disappointment when I had to tell them, as so often I did, that I was too busy for a game after tea. At least Bressingham was the kind of place where they had scope for romping and playing, and the fact of their being free, yet settled once more at home with me, was with my work a considerable antidote for the loneliness I would otherwise have felt more keenly. There was compensation too in the renewed contact with those helpers who had stayed on, who were glad to have the boss back on the job. At least I'd got something out of my system through going to Canada, wasteful though it had been in effort and money, because now there were no doubts left as to where my heart lay, along with my duty.

But contacts with steam were few and far between. On rare trips to London for meetings I felt I ought to attend, I'd take the train from Diss. Up to 1951 the locos had been Thompson's 4-6-0 B.I's, and though vastly superior in my eyes to the old G.E. 1500's they were often late back and laboured heavily up the long Brentwood gradient. Then came news that some brand-new locos were coming to replace them on the Norwich–London main line. I asked Don, who drove the lorry to Diss station with orders for plants to go by rail, to tell me if ever he saw one come through. Not long afterwards he sought me out on his return one early spring day in 1951. 'Saw one,' he announced brightly. I was busy counting plants into a box I was packing. 'One what, Don?' I asked unthinkingly. 'Why, one o' them new engines you told me to look out for—and they're called Britannia Pacifics, or something like that, but talk about big! They're painted green, with no end of wheels and the whistle sounds different to any I've ever heard before, like two notes run together and I rather like the sound.' I'd lost count right away and asked Don for more details. 'Dunno as I can tell you much more. They say these new engines come through Diss station at ninety when they don't stop there. The one I saw did stop, but by the time I walked up to the end of the platform, it were off again. I did notice it had a peculiar sort of shield fixed both sides in front that made it look bigger still. Cor! It were a size though!'

This was indeed exciting for me. But though too busy to break off during the spring rush season, I knew that next time a request to attend a horticultural committee came, I'd find less excuse not to go to London than was usual. A few days later the wind swung round to the east. From that direction one could often hear trains passing through or pulling out from Diss station, three miles away. Out on the open nursery I too heard that distinctive, musical whistle of one of the new Britannias and it sent a thrill down my spine, as I conjured up a picture of a great locomotive at speed.

The first Britannia I saw was when I went myself to Diss station one evening to take an urgent order for despatch. It was so majestic that it held me spellbound. True, there were certain features that were a trifle continental-looking, such as the smoke deflectors that Don had called shields; and the appearance of various outside tubes and pipes was also a break from the traditional smooth lines of older British locomotives. But to me it appealed instantaneously, as the embodiment of the noblest and grandest I could imagine or ever hope to see in a steam engine. It was a symphony in steel. Though lost in admiration, I thought how appropriate was the name 'Oliver Cromwell' set boldly on the smoke deflector. The namesake was an East Anglian himself, but whatever names the others bore, the fact that British Railways had built such engines and sent them to serve on East Anglian main lines brought a decided upsurge in my opinion of them. The whistle blew in response to the guard's green flag, with the driver leaning forward to raise the throttle lever. He must be a proud man, I mused, and with a thought flashing in that perhaps one day I might even be allowed a footplate ride. That thought was overwhelmed by the sudden and irresistible wave of emotion that swept over me, as those great wheels began to turn, forced into motion by the piston and gleaming connecting rods. My mental activity became blanked off until I came to, realizing that the train had gone and only the pennant of smoke remained above the coaches, to show where 'Oliver Cromwell' was charging his way through the Palgrave cutting, with gathering speed, towards London.

The twinges of sadness I felt as I returned home were nagging in their effect. The chances of obtaining permission to ride on the footplate must be pretty remote. Believing that such a privilege could be granted only to the most important or influential, I could scarcely imagine anything but a curt refusal if I were to ask. It would be better not to risk the hurt a refusal would bring, and just now I was finding how impersonal British Railways were to deal with. I'd been trying hard to obtain special rates for plants we despatched by rail, and interviews with officials concerned had been rather frustrating. They had their rules, it seemed, admitting of no deviations, and appeals to those with authority became blocked. At present, I was told, the amount of traffic we placed with B.R. did not entitle us to any substantial reductions in rates, which then were about ten per cent of the value of plants despatched. Besides, even if I were granted a footplate pass, might I not then feel even more acutely the lack of association with steam, following the loss of Bella? It might even prove unsettling, and with the uphill task I'd set myself just beginning to bring some reward, it would be unwise to take the edge off my determination.

Not long afterwards Tim Stevens called in at the nursery, his lean face all smiles. I'd seen scarcely anything of him since my return home and knew that with no engine here there was little incentive for him to call.

'Just thought I'd come and tell you I've bought an engine,' he said gleefully. With a wry smile I waited for more, since he was obviously bubbling over. 'And she's a beauty, a Burrell like yours was, only she's a single crank compound, one of the best types Burrells ever made.'

'Are you going to use it?' I inquired, trying hard to show interest rather than envy. 'I reckon so. I've got her in a shed beside my house and when I've done one or two little things to her, I'll get myself a saw bench, and amuse myself in my old age. Yes, she's a real beauty. Number 3193, about 1909 or 1910 I'd say.'

'Well, well,' was all I could say, but as he prattled on I made a few appropriate interjections. 'Yes, it was a crying shame about your engine. That was a good engine too, it was. And I s'pose you never did get paid for it, nor find out who they were that cut her up and took her away? Even if it had been me that let it go for scrap, I'd have the money first, that I would. Daylight robbery that were, and such people ought to be behind bars'—and so on, till I could stand no more.

'I suppose there's very few engines lying about now,' I said, not only to put an end to what I had no wish to hear.

'Very few, very few indeed. I was lucky to get mine, such a beauty she is, but these scrap boys they've nosed 'em pretty well all out now and for as little as £10 apiece. Why, my cousin sold three of the best engines you'd find anywhere for thirty-five quid, only three years ago that was. Why, are you thinking of trying to get another one then?'

'Just thinking,' I admitted ruefully.

'Well, if I do hear of one, I'll be sure to let you know. And if I don't and if the time comes when I decide to part' his voice fell to a whisper—'between you and me, my wife doesn't altogether approve, thinks I'm too old, though myself, I don't—if I do decide to part, then you shall have first refusal. How about that?'

I thanked him, saying that I badly wanted to get a replacement for Bella, but I'd have to wait perhaps another year or two. He shook hands with me and said, 'Well you may depend I shall keep that in mind. I know you'd give her a good home and give her some work. Engines are best worked, they are indeed, and now I must be off.'

I kept it well in mind too, for it seemed that Tim might well be persuaded, if not obliged, to part within a year or two. By then I ought to be able to run to one, thinking that £100 or so would buy it. Besides, such an arrangement, though by no means concrete, was something to hang on to, as a hope that would anchor down to a large extent my hankerings which of late had become much more lively. The trouble about owning an engine, something which bothered me as much as the cost, was that I feared I should not resist the temptation to have it in steam, with no good excuse other than my own pleasure. With a more powerful tractor now in use, it could scarcely be justified as a piece of useful equipment now that no more economically worthwhile reclamation projects could be considered.

7
On the Boil

Looking back over the years, a peculiar fact emerges. For most of the 1950's I had to push steam into the background. It was no use crying over spilt milk, nor could I dismiss for ever my hopes of renewed ownership. But I can see now how the austerities of those years paved the way, although I could not foresee this, for the swift developments of the sixties. Having to forgo or defer the pleasure of owning even just one engine, the thought of building up a collection did not enter my head, and it was a sequence of events connected purely with the farm and nursery which opened the door to steam. Had I allowed a determination to buy another engine to become fixed, it might well have taken priority over other projects and prevented the unfolding of later opportunities.

These arose as a more natural corollary to the nursery business, situated as it is as part of a farm. This must therefore be the place briefly to fill in some more background. By 1952 it was pretty clear that the nursery business was pulling round and that it must grow beyond its pre-war size. Every penny that could be made and spared had to be ploughed back, and with expansionist ideas came the need to utilize fully every usable acre. More mechanization on both farm and nursery, more buildings, better irrigation and more staff boosted production to meet increasing sales, but because this was progress I worried much less when the bank account regularly plunged into the red during the slack period for orders. That period—May to October—was the one in which improvements had to be made, but I had no doubt that eventual, though modest, prosperity would result.

Through past experience I knew that fat profits were unattainable. The element of chance on both the production and sales side was far too great. Stock failures through weather vagaries and a quite unpredictable demand are constant bugbears, yet on balance a reasonable margin could be expected.

A nurseryman's life is an exacting one, if he leads it conscientiously. If he has also a love for plants, it can be very absorbing too. To rear plants from cuttings, seed or division and to handle them in their thousands up to the time when a year or two later they are fit for sale, can generate a feverish activity, an interest you cannot easily let go. Different kinds have to be dealt with, propagated, transplanted by several different methods at the correct time and grown on in correct conditions of soil, situation and moisture. Knowing that even greater variety exists somewhere or other, one is instinctively on the look out for additions to the range already in stock. In hopes of enhancing that range and stimulating wider sales, the temptation to bring in more is hard to resist. Many kinds have potential for breeding new varieties, and this is a lure I've always felt. But for every new variety I introduced that is worthy of being launched as a novelty under my choice of a name, hundreds have to be discarded. And their selection and re-selection, spread as it mostly is over a period of years, is as time-taking as it is absorbing.

By 1952 the nursery acreage was back to its pre-war size and there would be more than thirty-six acres under plants by 1953. The peaty fen fields on which I'd set high hopes were proving disappointing. The soil was not so fertile as it looked, and waterlogging was much too frequent to make either nursery or arable farming a reliable proposition. There were snags too on the loamier fields on the slope above the valley bottom. The soil was prone to panning, becoming hard when dry and sticky when wet so that either way it was difficult to deal with when plants had to be dug out or set. They grew well enough, but the clinging soil added considerably to freight costs. Between the two types of soil was a strip of lighter sandy loam. This was ideal for plant growing but there was too little of it, and so by 1953 I was looking for land again, despite having two hundred and twenty acres, all told.

Profiting again from hindsight, it is all too plain now that had I pledged everything to buy land at that time I should have been quids in. From £60–£70 per acre then it has jumped to £300–£350 per acre now. I bought thirteen acres because they were next door, but because it was still stiffish land I hired some sandier stuff a mile away, which proved inconvenient. Then came the idea of improving the two best fen fields, which were conveniently situated close to the centre of things. In one was a sandbank, and whereas the black peaty parts had sunk through drainage and cultivation, the sand had not. By taking out sand from beneath the low banks, and spreading and mixing this in with the peat, I thought a double benefit would accrue. The mineral sand should help the peat and the levelling off bring both greater fertility and stability. This is what I believed. It worked to a fair extent, but it was a costly process. The field adjoining had no sand bank to go at and it lay wetter for longer periods. One winter we had to dig out by hand two acres of phlox plants which were lying, as a result of flood and frost, beneath a low roof of ice. The remedy of added mineral soil, which in turn raised the level, could only be applied during a dry summer.

Such a summer came in 1955 and with an eye to landscaping, and with a liking for punting, fishing and skating, we dug an extension to the big pond in the field above. Using our own tractors with rear bucket scoops, we took out four to five thousand cubic yards of sandy clay and spread it where we thought it would do most good. It took five weeks and at the end we'd made the pond extend to two acres varying from two to eight feet in depth, to be kept full by natural seepage.

This project was in itself quite satisfying, for I enjoyed the labour entailed, and another year passed with steam well in the background. Things were thus in 1956, when at last I became free to marry again. And Flora, apart from other blessings she brought, was the kind of person to encourage further ideas of making improvements, including some of a more aesthetic nature. I'd had sixteen years in the marital wilderness, and with domestic harmony at last, incentive was not lacking. By now, Bridget was emerging with honours from University, Robert had finished school and had left home to study farming, Adrian was in his last school year and wanted then to see the world. I could scarcely complain that neither boy had leanings towards the nursery as a profession. They had seen and to some extent suffered from the exactions it imposed on me, and they must be free to choose their own way in life.

By 1957 the time had come and the opportunity was ripening to carry out one of the nebulous ideas I'd had on first coming to Bressingham ten years before. I'd made something of a garden in front of the house on what had been a large untidy lawn backed with a shelter belt of trees and evergreens. These beds had been something of an experiment in 1952–5, used as a means of growing hardy plants and breaking with the conventional plan of a one-sided 'herbaceous border'. The results in terms of effect and ease of maintenance exceeded my expectations, and I decided to extend the garden on similar lines. Beyond the shelter belt lay eight acres of park-like meadow, with pleasing undulations and some fine mature trees. The soil was good and all I had to do in order to be able to grow a much wider range of plants was to carve out beds in keeping with the informal surroundings. It seemed such an easy way of satisfying my interest in plants, and at the same time of bringing some indirect benefit to the business. Not only could reserve stocks be grown there, showing prospective customers how they grew naturally, but I could plant some uncommon kinds which, after trial, might enhance the range offered for sale. Sales were effected almost entirely by means of an annual catalogue, available only to the trade, and retail business had been rigidly avoided since 1934.

It was a case of finding myself caught up in an idea. The taking over of an acre of meadow for the new garden in 1957 was, as I said, an experiment. But before long Flora and I began a series of visits to gardens both private and botanic, in Britain and on the Continent, and swapping brought in a seemingly endless variety of little-known species of plants. With more help on the detailed management side of the business, I soon found myself drawn to further extensions, and as fresh kinds arrived for planting, more beds were dug for them. It was fascinating work, and when the need for retaining walls or a shelter arose, it came as a pleasurable relaxation to build them myself with local flint stones.

After the first winter's work of making new beds, doing some walling, and constructing two pools linked with a little stream for moisture-loving plants, I found myself planning all summer long the shape of the next intake. I enjoyed both the digging and the planting as well as the propagating needed to increase a single plant of some choice species, making from it a group of a dozen or more to achieve the best effect.

To begin with, it had never been a part of this project to allow the new garden to be open to public view. There was no wish to stimulate demand for plants from amateur gardeners, yet the time came when we had to let them see the new garden. Before 1958, requests from a few charitable organizations had led to the nursery being open occasionally during the summer. With voluntary helpers, a Sunday afternoon might benefit the Gardeners' Benevolent Fund or the National Gardens Scheme for the nurses to the tune of £20–£30 if the weather was fine. But the new garden near the house was taboo to outsiders. It was something private, and apart from a few coach parties of horticultural societies, we kept it for ourselves and for the increasing number of gardening friends we made on our journeyings in search of plants.

But as the garden grew in area and variety, so the pressure from would-be visitors increased, and in 1958, following a B.B.C. featuring of the garden in a television programme we had to give in. Once we allowed people in the garden, even if only on one Sunday afternoon every month during summer, it was inevitable that more frequent openings must follow. It was, after all, much pleasanter for visitors to see plants flowering under garden conditions rather than to wander up and down nursery fields. So while still holding out against retail selling, we saw that the increasing number of visitors offered scope for raising money for charitable causes, which was in itself worth the small cost and inconvenience to us.

With precious little idea of how this aspect would snowball along, once accepted as a worthwhile principle, I found there was added zest in making the garden even larger and more complete. Having come to realize that there was no large garden in Britain where hardy perennial and alpine plants were the principal feature, I found this to be an added spur. Everything, or nearly everything, was favourable from the point of view of soil suitability and situation, and it was fitting in very well as an asset to the nursery business. As well as giving me an outlet as a plantsman, being able to design and make a garden according to my own fancies was a joy in itself. And if people wanted to come in increasing numbers and see it on Sunday afternoons, it could be taken as a compliment and we could hand over what they paid to come in to the causes we thought most deserving. To the first acre taken in as new garden in 1957, each year till 1961 brought an addition of about one more acre. Undulations, and wide grass paths surrounding informal island beds, became the established pattern. By 1961 nearly five thousand different species and varieties of plants were growing in the beds.

Even so, by 1961 I'd seen the red light and the impetus slackened. It was not difficult to keep making additions in area and content of the garden, carved from rich old meadow sward. Weeds were few at first, but when it became clear that maintenance would be an increasing problem the larger the garden grew, it was time to call a halt. Besides having to leave room for visitors' cars to be parked on open days, I had to consider the demands made on my time. Though helpers had been brought in, I was the only one with the knowledge and experience to ensure that the beds were properly maintained. It was not merely a case of planting and keeping free from weeds. With such a wide variety, placing, propagating and extra care were necessary for this kind of garden, with its emphasis on choice or uncommon kinds. The red light indicated not only that expansion must cease, but also that my mind was becoming a kind of filing system for innumerable plants each represented by its botanic name. I needed an antidote, and seeing the need I at once brought steam engines back to the forefront of my mind.

Nine years had passed since I'd classified my love of steam, as a hankering that might one day be allowed to emerge but not yet. It had in fact come to the surface more than once during that time only to be forced back again for one reason or another. Three or four years after Tim Stevens had called to say he'd bought an engine, I heard a rumour that he'd sold it again. It made me feel sore, so sore that I nursed a grudge while resisting the temptation to tell him about it. In time this resentment eased, especially when I also heard he had sold it to Leslie Aldrich, whose father had bought the engine new from Burrells when Leslie was a lad in 1910. That was a reunion I could sympathize with, but it would have helped had Tim explained at the time. Following that discovery I made a few tentative inquiries and learned that not only had traction engines become even thinner on the ground, but that a movement towards preservation of those remaining had begun.

Once or twice when on a journey I had seen a notice of a Traction Engine Rally nailed to a wayside tree. This was stimulating in a way, but the fear came that if an engine for sale should be located, it might be too pricey for me; just at that time I'd bought another seventy acres of land. All the same, I let it be known to one or two visiting agricultural reps., who might be in the know, that they could tip me off if they heard of one that might be had.

I was working in the garden one day when a thick-set, red-faced man came up and after making the usual observations said he'd heard I was interested in buying an engine. I straightened my back immediately and took notice. 'I've got two and I wouldn't mind parting with one,' he said, 'a nice little Garrett Tractor. Doesn't need a lot doing to it.'

'What's the figure?' I asked. In response to this he went on to explain that there were very few about now. He'd searched all over Norfolk and a good deal of Suffolk because it was evident that traction engines were going to be wanted. Clubs and societies were being formed to promote rallies which were being attended by lots of people. So he had bought a few and soon found buyers for them, though he had retained two or three as his special pets and had fully restored them to their former glory. If I was in a mind to buy, he added, then I'd best not delay, because the increasing demand would make them harder and dearer to obtain and they were sure to go up in price.

It came as a surprise that the demand, if what he said was true, was sufficient for him to become a dealer in traction engines. He had not described himself as such, but in so many words had admitted it, and this knowledge rather put me against him. Somehow it didn't seem right that anyone should buy steam engines nowadays in order to sell them again at a profit. He told me his name, and it was one I'd heard in conversation; he had been once or twice referred to as one whose interest went beyond ownership. 'What's the price?' I repeated, though it is customary in East Anglia to hold back mention of the price till the very last when trying to do a deal. 'Well,' he said, 'seeing as it wants a bit doing to it—nothing very much of course, and you can come and see it any time—it's pretty cheap at four hundred. Garrett's tractors are nippy as you no doubt know—just right for taking to Rallies.'

I cannot be sure whether that was in 1957 or 1958, but confronted with such a figure, it was a case of having to put the notion of buying an engine farther back in my mind than ever. I just hadn't that kind of money available, either from the business or personal account.

8

In Steam Again

Flora would take it as a dubious tribute now that I compliment her on being the one to persuade me to indulge once more my longing for a steam engine. Her response to that would be that she never dreamed it would result in my having more than one. What she saw perhaps more clearly than I, during the winter of 1960–1, was that the garden was becoming too much of an obsession with me. It did not end with the day's work in the garden, and keeping tabs on the running of the farm and nursery. I'd been asked to write articles and books on my branch of horticulture, as well as to go on television and radio programmes. With the subject fresh in mind, articles especially were easy enough to write, and the more I wrote, the more various outlets opened. Being very unwilling to give up precious daylight hours, I had to write my articles and books after dark, mostly on winter evenings. These at least brought a boost to my private account at the Bank, which for years had languished for lack of support.

Some friends whom we'd not seen for several weeks called one evening in January 1961. Once seated, John fished out a piece of paper from his pocket and handed it to me with, 'Sorry I've been such a long time getting a few likely names and addresses for you.' And then I remembered. John was in the agricultural machinery business, and the last time we'd met I'd told him I'd like to get hold of a traction engine. He'd said, as others had who were similarly placed, that he'd make some inquiries and let me know. And here was proof that he at any rate had taken it seriously, even if it had taken time. There were only three names on the piece of paper, and one of them I noticed began with 'Rev.' Surprised, I read it out—'Rev. R. Stebbing, The Rectory, Tacolneston.' 'From what I gather', John explained, 'he's reckoned to be an authority on traction engines. He's President of the East Anglian Traction Engine Club, and he's apparently as good a bet as any to put you on to someone who might be willing to sell.'

Then I also remembered hearing or reading about a parson who had bought an engine and was driving it home when its rear axle snapped. Rather than cause an obstruction on the road, he'd been persuaded to send for a scrap merchant to cut it up on the spot and cart it away. I'd felt sorry for him, but even sorrier that when engines had become so scarce one should have been destroyed in such a way. 'Why don't you give him a ring?' suggested Flora, when our friends had gone. I did, but there was no answer. Next day I tried again and Mr. Stebbing said he'd ask someone he knew who had decided to sell. 'I'll ring you back,' he said, 'it's just possible he hasn't yet sold it.' Half an hour later his voice came over the telephone again to say that although a man had said he'd buy this engine, a month had passed and the owner was prepared to do a deal with someone else if he could be sure it would go to a good home. That was on a Friday evening. Flora suggested that we go to see it on Saturday afternoon, saying that it would make a change to drive over to Sudbury, where the owner lived.

When a hankering had been stifled for years, it's apt to become a habit to treat it as a will-o'-the-wisp, or a pie-in-the-sky. To discover suddenly that it can be pandered to, that there is no valid excuse for not grasping opportunity with both hands, calls for some mental adjustment. That was where Flora's influence came in. 'I'd say it's just what you need,' she said. Being a farmer's daughter she added that for her, too, a traction engine was a fascinating piece of machinery, and told me how excited she used to become when a steam-powered threshing set used to winch itself up the hill to her father's Yorkshire farm. 'I'd love to see one here,' she went on, 'and I might even have a go myself at driving it after the babe has arrived. Come on—let's go tomorrow afternoon, and I'm sure you won't regret buying it. It may not be all that expensive, and you've got some writing money which you ought to spend on something you want for yourself for once, instead of putting it back into the business.'

There was only one answer to such talk. It was all I needed to bring back a flood of yearnings that were quite irresistible. I made an appointment by telephone with Mr. Gerald Dixon, the owner, to go next day. Knowing full well that the die was cast, I could scarcely contain myself for excitement, but just as a precaution I also telephoned next morning to the two other names John had given. It was as well to find out in advance all I could about current values of traction engines, so as to be able to judge whether or not Mr. Dixon's figure, when the time came, was in fact a reasonable one.

After we had knocked at his door, some time elapsed before we set off again to see the engine. Mr. Dixon, who was much like Tim Stevens in age and build—short and slim—like Tim had spent a lifetime with engines until he retired. After retirement he wanted to own an engine as a hobby; he had found one discarded in a bed of farmyard nettles, and then had become a prime mover in founding the East Anglian Traction Engine Club. He showed us engine photographs over a cup of tea and talked of engines while popping the odd question to me as to why I wanted one. My answers, including an assurance that I was not a dealer, seemed to strike the right note, and then we drove a few miles towards Braintree, with him as guide, till finally we pulled into a yard. 'There she is,' said Mr. Dixon, and as we took off the canvas tilt, he explained how he and a friend had stripped her down to change rust into brightly shining steel and dark green paint.

Of course we'd been told before seeing the engine that it was a single-crank compound Burrell, weighing about eleven tons, and that it was built in 1909. It was, in fact, exactly the same model (and only a few months older) as the one Tim Stevens had bought and sold again to Leslie Aldrich, the only difference being that it had a cast-iron instead of a copper-topped chimney. Mr. Dixon apologized for its not being in steam, but I was welcome, he said, to come again another day and try her out in steam if I wished, and could spare the extra time. But as I walked round and round, I was so taken up with the thought of ownership that I had to hold myself back from asking the inevitable question too quickly. I was, in fact, having to put on a bit of an act, partly to cover up my lack of knowledge, and also because Mr. Dixon might have thought it odd to rush in at once and ask the price before a decent interval for examination had elapsed. I was convinced that what he told me about its condition was true. It was in good order now, and he was only selling because a recent illness and increasing age were making it too difficult to retain ownership. 'And the price?' I asked, making it sound casual. Just as casually he replied, 'I think a hundred and eighty would be about right. I'm not out to make money, and because I can't look after it as I'd like to, I just want to make sure someone has it who will.'

On our way home again Flora said, 'It didn't take you long to say yes, did it, or to write out a cheque? You didn't even tell me you had your cheque book with you.' I laughed, happy as a dog with two tails, and joked about the other man—the chap who had taken a month, and still hadn't made up his mind whether to buy or not.

The first job I tackled next morning was to make a place in which 'Bertha' would stand. She must of course be given a name. 'Bella' was the first one to occur, but on reflection sentiment ruled it out. Bertha was just different enough as a name, and ran well both with 'Bloom' and 'Bressingham'. On that Sunday morning, instead of working in the garden I levelled off a piece of rough grass and spread some gravel in the lee of four pine trees halfway between my house and the nursery office. It was an ideal spot, because I would pass within touching distance a dozen times a day. Late on the Monday afternoon, after hours of excited anticipation, the low loader drew in and there stood Bertha in its well, securely chained and chocked. With the loader's rear wheels out, a tractor was needed to pull her off on to the standing I'd prepared. She was here, she was mine, and tomorrow would, all being well, see her in steam and motion.

This knowledge banished any lingering animosity there might be left towards Tim Stevens. Maybe he'd never suspected I had any, and had forgotten he'd promised me first offer of the engine he had. I was certainly not going to remind him when I told him over the telephone that evening what was now standing in the yard, and asked if he would like to see it in steam next day?

Tim arrived as she was beginning to make steam. For the next half hour, he gave me another unsolicited lecture on care and maintenance. Most of the advice I remembered from Bella's days, fourteen years before, but he came out with so many 'do's and don't's' that by the time she was ready to move, my old feelings of awe had returned. The joy of renewed possession gave way, to some extent, to fears of making some blunder which might be damaging to my pride as well as to the engine. Half of me wished that neither Tim nor anyone else was present so that without witnesses I could again get the feel of steam in motion, while quietly learning to master her. But the other half needed a refresher course to achieve that mastery without mishap.

Tim's guidance was so persistent that it became obvious he was regarding it as a treat as well. 'Let me show you' came with what I thought needless frequency, and though I moved over each time to let him take the controls, it was a relief to hear him say it was a pity he couldn't come back after lunch. All we did was to chug round the yard, backing and shunting round the buildings. With a compound, the steam from one cylinder passed on exhaust to the next, but on this type of engine the pistons were so arranged that both drove the same crank. There were two speeds, with a simple but clever device for changing gear, though this had to be done with the 'motion' still. Though a little more complicated than the single-cylinder engine, Bella for instance, for all that 'Bertha' was fifty-two years of age her gears, bearings, and the whole motion worked smoothly. If out of gear, the cranks and fly wheel alone could be made to turn so slowly, by cutting down the throttle, as to revolve only about ten times a minute. And not a sound could be heard. 'Yes, you've got a good 'un,' Tim remarked as he finally climbed down from the footplate, 'and it makes me wish I'd never sold mine.'

Though of necessity the business claimed my time that afternoon, I slunk back a time or two to keep Bertha's fire alight with a few chunks of wood, since I had no coal. After tea, I spent another half hour with her, with Flora at the steering wheel. Then, as darkness fell, I drew over the canvas tilt but did not tie it down, knowing I'd have to come out once more before leaving her for the night. When bedtime came, I flashed a torch and saw that still a few pounds' pressure were registering on the gauge. The warmth and the smell of oil and steam, so evocative to so many, made me fold back the cover, and I opened the throttle and watched the motion in the light of the torch until no pressure at all was left.

The inexorable, exacting rush of spring work in nursery and garden brought scarcely any opportunities of steaming Bertha during the next few weeks. Even when passing her at the double it gave me a thrill to know that at last I owned an engine again. But no sooner was spring turning to summer than a thrill of a very different kind came my way. I had seen the name G. G. Goodings, Divisional Manager, printed at the top of correspondence we had been receiving from the Norwich division of British Railways, over the old question of freight rates. The organization had of late become easier to deal with as a result of more negotiating power being given to area officials. Our account with B.R. had increased, encouraged by special rates agreed with the Norwich Division, but I'd had no direct contact with Mr. Goodings personally until one day he telephoned to say that he and his wife would like to pay us a visit, partly on business and partly because he had a large garden in need of some plants. Prompted by the opportunist side of my nature I asked them in to tea. With conversation ranging from Begonias to Beeching and Britannias, over tea, I plucked up courage to ask what had been at the back of my mind for donkeys' years, awaiting such a chance as this. 'I suppose it wouldn't be possible for me to have a footplate ride?' I asked haltingly.

'On a Britannia? Why, yes, I think that can be arranged. But you mustn't leave it long, because we've begun to lose them. The diesels are taking over in East Anglia, and I doubt if steam will last out the year.' Good news and bad news all at once, I thought, but the idea of having a footplate ride at last was uppermost, and I could scarcely believe it. Though it had been a shock to learn that the Britannias were leaving East Anglia, it had recently been reported that steam would be phased out so slowly that it might well last here and there till the end of the century.

'You may find it a rough ride if you've never ridden footplate before,' went on Mr. Goodings, taking out a notebook. 'When would you like to go? You see, I must arrange with an Inspector to go with you. That's regulations. Right you are. May 31st. The 7.45 Norwich to London, calling at Diss at 8.02—and take my advice, don't put your Sunday best suit on. I'll get Inspector Sands to travel with you. He's steam-minded, and you'll get on well with him.'

It had been so very easy that it made me wonder if, after all, I might have been allowed the privilege years before, had I dared to ask. But when the day came, any regrets on that score vanished as in the sunlight of early morning a fleck of smoke appeared in the distance. Flora had brought me to Diss station, but as we stood at the end slope of the up platform, I could think of nothing to say. The great locomotive thundered in, to stop just clear of the platform. Inspector Sands, guessing that the chap in overalls must be me, beckoned me to climb up. From ground level it seemed a very long way. Flora stood down there with the camera, obviously baffled by the size and close proximity. There was a look in her eye that said she wished she could come too, but was happy so long as my dream had come true.

It was not until I'd had to stop waving from over the cab door, having crossed the bridge over Victoria Road, that realization came. Yet though I tried to tell myself that I must not miss a moment of this thrilling experience, I found emotion came welling up to dim my wits. Mr. Sands smiled and made a few friendly comments, but I could scarcely find a rejoinder and in any case the noise was reaching crescendo. The rattles and vibration were far greater than I'd imagined. So was the swaying, and the sight of the massive length of boiler in front rolling from side to side was frightening to watch for someone who had never before been on an engine at speed. The fireman began to stoke, and then offered me the shovel. 'Like to try your hand?' he shouted with a grin. The enormous furnace glowed white hot with dancing flame. Not wishing to appear a jibber I tried my hand, but after two vain attempts to imitate his deft movements, had to shake my head and give back the shovel. It was more than just the knack I lacked, nor was it only the swaying of the engine that affected me. It was, as much as anything, the feeling of utter bewilderment which would not leave me, making me only half conscious that I was there at all. Yet the knowledge—the evidence—that indeed I was, was almost painful in its intensity.

By the time we reached Ipswich such feelings were easing a little. 'She's a rough rider,' commented Inspector Sands. I told him that I'd been so excited at Diss that I'd not noticed the name. He laughed. 'It's 70034, "Thomas Hardy", and for your sake I could have wished it had been "Oliver Cromwell" or "Hereward the Wake"—they're better riders. Still, you're getting used to it, aren't you?' I nodded as we steamed into Ipswich tunnel, but by the time we reached the sharp curve between Brantham and Manningtree, over the estuary bridges, I became a little scared at the speed we were making. From Colchester onwards there came more of the unalloyed enjoyment I'd expected of the whole trip. The skill of the driver as his hands automatically manipulated the controls filled me with admiration, as did the almost casual attention of the fireman to the injectors. The co-ordination between driver and fireman was perfect, without a word being spoken. The latter so obviously knew what was expected of him, and did it, and there was plenty to do. Such trust was as obvious as it was vital, as was the trust which, as the great locomotive ran over each maze of points, showed that all were set clear by the signalmen.

Then we were within the dim recesses of Liverpool Street, and the ninety-five-mile journey was over. Inspector Sands and I talked for an hour over a snack. He spoke proudly of his son Geoffrey, who had just been appointed locomotive shed master at Crewe, before I had to leave him for an appointment with publishers. I'd had my fill for one day, and on the way home in the comfort of a compartment, fell to wondering how soon I could decently ask for another footplate ride. If only steam could stay on long enough I'd try for a night journey next time, and perhaps a trip on a goods train as well.

9
'Any Old Iron'

The feeling of having, as it were, gained a foothold in steam was uplifting. Apart from adding a new dimension to my interests, I felt there was a vague aura attached to it, at least at that time. On Whit Monday in that year of grace, 1961, Flora and I went for the first time to a steam engine rally, taking six-weeks-old Anthea in her carry-cot. This was the third annual rally of the Norfolk Steam Engine Club, held at Woodton Old Hall, the home of the club's Chairman, George Howlett. The Secretary, Cecil Knights, gently chided me for not taking my own engine to the rally, and I wished I had, but promised to do so next year. What surprised me was the enthusiasm of so many people for steam. Gate receipts had steadily mounted and this year over seven thousand people had paid to see thirty road and traction engines on show. Standing at first in orderly formation, ticking over for public inspection, the engines next moved in a massed parade round a ring of several acres. Then there were obstacle races, hooking the bale, musical chairs, a slow race and ladies' steering race, all with engines with belching smoke and their works gleaming in the sun.

It was a brave sight. Here was proof enough that the days of steam were not over, even if the economic age of steam had passed. And if nine out of ten of the engines for farm, haulage and road rolling had been cut up for scrap, enough remained to foster and give vent to the nostalgia and fascination for steam engines that lay in the hearts of so many. In East Anglia steam had been used for farming and haulage probably to a greater extent than anywhere else in Britain. Judging by the number of engines here, it appeared as if even in their obsolescence, the East Anglian heritage was rich. The ornate Showman's type, with its twisted brass uprights holding the long canopy, was represented, and so were the big ploughing engines. The Rev. Mr. Stebbing, who had put me on to Bertha, had recently bought one of these, a Fowler weighing twenty-one tons, and had driven it over in spite of some faults which he'd had no time to put right. As I was chatting with him, someone came up and called out, 'Where's her sister?', adding in the same bantering tone that one ploughing engine by itself was only half the pair needed for work. Mr. Stebbing countered by saying that he was a bachelor, and Cissie was as much as he could cope with, without having to look after Bessie as well. To me, standing beside him on the footplate, he added, 'I know where Bessie is, if you're interested. She's in Duce's scrapyard at Cambridge.' I shook my head. 'Thanks all the same, but I think one engine will have to be enough for me,' I said, believing this to be the case.

The Sunday following that Whit Monday was the first open day in 1961 for the garden. From once a month, openings had increased in 1960 to every other Sunday from June to October. This was not a very good compromise, because not everyone who heard about the gardens at odd times on radio or television saw the press advertisements giving opening dates and times. Most of us were against the idea of having visitors invade the garden every Sunday, and though this seemed inevitable sooner or later, we decided to stave it off for another year.

The increase in numbers of visitors had also involved this change; sometimes we'd provided staff for parking and gate duty ourselves, now we passed it all over to volunteer contingents from local organizations. The St. John's Ambulance Brigade, Round Table and Rotary Clubs and Diss Horticultural Society took turns, and in return were to be credited with a share in the takings for distribution to their own or any appropriate causes. Gate receipts of £40 on a fine-weather Sunday at 1/6d a head for adults was considered good, and the system of a final share out was much fairer than the former method of holding open days for specific causes. The weather could make or mar efforts on those lines, and there had been occasions when receipts had not covered the cost of advertising.

On that first open Sunday, Bertha was left covered over to begin with, since I was patrolling the garden. Later in the afternoon I spotted people peering under her tilt, and in response to a request took it off. After this, other people came and there I stayed for an hour or more, wishing I'd steamed her since visitors found her so interesting. A fortnight later she was in steam by the time the gates opened at 1.30. It suited me well enough, because it gave an excuse to stay, and a worthwhile one at that. It suited me much better than sauntering round the garden answering questions from visitors, and keeping an eye on would-be pilferers. Someone else could do that, and being in charge of Bertha was much more of a change. In spite of the crowd at Woodton Rally it had taken me by surprise to find that so many of those who came to see the garden paused to see a traction engine. On first thoughts the two scarcely seemed to be complementary, but my second thoughts seemed to show they were very much so. It was something different to watch the pleasure of those members of a family party who, though not specially interested in gardens, had nevertheless been persuaded to come. And from remarks I overheard, it was obvious that a new generation was emerging which included numbers who had never seen a real steam engine, only pictures of them.

Standing there, with Bertha gently ticking over, I could look down from the footplate and see a score of people around, just watching the smooth rhythmic motion. It was a study to interpret their expressions. The countryman of mature years was thinking of the old days, when engines were so much part of East Anglian farming life, and for him it was like seeing an old acquaintance once more. His wife would nudge him and ask a question, which set him off explaining how the thing worked, or recounting some anecdote in which an engine was involved. The younger man or woman too, whom I judged to be from a town, would also stand and stare. It was more difficult to guess what their thoughts would be, but even if neither evocative nor nostalgic, the fascination of outsize moving parts, set alive by steam, was quite enough to hold their gaze.

It was the children, both girls and boys, whose faces I liked best to study. Small boys especially gave away their feelings as with heads held back they looked up in wonderment and awe—just as I had done forty or more years before. I had greater opportunities then, for steam was still in its heyday. Yet no one ever had a steam engine on view like this, because they were always on the work for which they were intended, and boys were able to watch only on sufferance. Following such a train of thought, I began to put myself in those youngsters' shoes, to feel again the thrill and elation when the driver said, 'Come on up'; to me that had been like a Royal Command, and now it was I who was in a position to give it. Those old engine drivers did not have to add—'If you'd like to', nor, 'Sit back there on the coal and keep still'. The look on my face made the first qualification unnecessary, and I knew well enough that only if I kept still and out of his way, would this or any future invitation hold good. The only motion I ever felt was the gentle rocking back and forth in time with the thrusting piston, with the engine driving a threshing drum. Boys were certainly not allowed to ride when the engine itself was travelling.

'Come on up,' I said to a couple of boys with a longing expression on their faces. They had stood there for several minutes already, early on that first Sunday afternoon that Bertha was in steam. 'One at a time,' I added as they both made for the steps at once. I was still burning wood, having more than we needed for the house, and as I arranged the logs in the coal bunker to make do for seats, a girl sidled up and smilingly asked if there was room for another. I nodded and held out my hand to help her up; the three of them squatted, proudly smiling to parents standing below. With over three acres of meadow for parking, there was no reason to keep Bertha just ticking over. The little group of onlookers would like to see her in motion: the children would have the extra delight of a ride, but none would suspect that my pleasure would be even greater than theirs, since I was both owner and driver of this noble old engine. Sliding in the gears, with hand on the regulator lever, I turned round and said, 'Hold on tight' to my passengers, gave a pip on the whistle and chugged round in a circle over the grass. It took less than five minutes and by the time the children climbed down, others were running forward with the 'Can I have a ride too?' look on their faces.

'I've been thinking,' I announced to Flora a day or two later. Flora was coming to know that this was my usual approach to some project which I could keep to myself no longer. 'What this time?' she queried. Often as not my mind was already made up, but it helped to talk it over, hoping for and usually obtaining approval tinged with warnings not to take on too much, or hearing a point of view which in my enthusiasm I had overlooked or played down. She too had noticed the interest Bertha had been to visitors, and had been glad of the chance it had given to provide me with a complete change from the world of plants. But when I told her what was in my mind, she was not so sure it would follow that if we got a second engine there would be that much more pleasure for me, or much added interest for the visitors. Then when I added that I still had a few hundreds in my own bank account, and that if I could find an engine in need of restoration it would not only give me a lot of satisfaction to bring it back to its old shine and polish and working condition, but would prove a good investment as well, she gradually found herself agreeing with me. 'But where can you find another?' she asked. Then I had to confess to having telephoned, during the day, to a Mr. Reg. Palmer of West Dereham. His was one of the addresses John had provided some months before. Mr. Palmer had said he had one or two engines still for sale, and suggested I went to look them over, if I was interested in buying.

Within hours we were on our way with Anthea on the back seat in her carry-cot. West Dereham is a long way from the market town of East Dereham, and when we finally reached the hamlet, on the edge of the Fens near Stoke Ferry, the road beyond seemed lost in the mountains of earth being heaved out for a new drainage scheme. Once beyond, Mr. Palmer's yard, as he'd called it, was plain to see in that flat featureless landscape. It was in fact a field, and so chock-a-block with what seemed to be junk, that even the entrance was tricky. Old threshing drums, elevators, excavators, piles of timber, huts and roofing with nettles abounding made it a sorry sight. But between and above, conspicuously for me, rose three or four steam engine chimneys. A large yapping dog greeted us menacingly, and then an elderly man emerged from a shed, sizing us up fiercely, it seemed. I remembered seeing him at Woodton Rally, smartly dressed with bowler hat and shooting stick, but he looked very different now. I told him my name.

'Ah,' he began, 'you're the man who rang me up this morning! Well of course you're a bit late. I've very few left now, but I've had plenty in me time. Sent a whole train load away one time, all cut up for scrap—twenty-eight engines, all at once. And now blow me if I don't wish I'd hung on, there's several people after 'em now. Sold one last week, and that Foster standing over there will be going to Bedfordshire as soon as I've put some fresh tubes in her.'

There was more besides, as we walked slowly among the incredibly varied items of rusting material, mostly of agricultural origin. Mr. Palmer's voice, well punctuated with swear words, was no less forbidding than his face, and yet conversation was easy enough. He was a Fenman all right, I thought, as he told me he'd been a threshing contractor for most of his life, and had taken on road works as well: the kind of man who, having had to struggle to begin with, could not stop working; and when steam engines gave way to tractors after 1945, he took up dealing and scrap selling as a sideline, if not a hobby. He reckoned he'd bought up and scrapped, for all his fondness for steam, more obsolete engines than anyone in Norfolk. But having destroyed so many, now he was seeking for any that might have been overlooked, with a view to profiting from the resurgent interest in steam. One could not help wondering what conflict of mind was his, because his own love of steam had been so much at variance with his business instincts.

With Flora having to stay in the car with Anthea, because the going was too rough for the folding pram, I went with Mr. Palmer to see a Burrell and a Ruston engine as well as the Foster he was re-tubing for its new owner in Bedfordshire. The last was much less dilapidated than the others. Then we made almost a circuit of the field, and came to three Burrell steam rollers, one behind the other, where they appeared to have stood for years. A roller would be an interesting addition, and when told that these had come from Doran Brothers of Thetford, who had a sizeable fleet in good working order before they closed down their business, my hopes began to soar. One, with the boiler lagging falling apart from exposure, was a single-cylinder eight-tonner; another, with a dilapidated canopy, was a double-crank compound of twelve tons; while the third would have been a ten-tonner if its roller had been there—but it wasn't. 'No, I sold that off it a year or two back,' Mr. Palmer explained, 'but you could put a pair of iron drum wheels on instead, because it's a convertible. That front saddle comes off easy enough, and with a two-wheeled fore-carriage, you've got a steam tractor. They're three good engines, they are—nothing wrong with the boiler or firebox of any of them, and I'll sell you one, or better still, I'll sell you the three.' 'What's the price?' There was no humming and hurring about Mr. Palmer, as he rapped out the prices he'd take for each one, with a discount if I took all three. Obviously this was a proposition to which further thought must be given. The temptation to do a package deal on the spot was so strong that I had to make excuses to be off, leaving Mr. Palmer with the promise to telephone him next day.

The first of the three rollers arrived three days later. It had not been difficult to come to terms with Mr. Palmer, and having taken Jack Clements with me as adviser, I felt as sure as could be that the best bargain lay in taking all three. Jack was known far and wide as being the greatest steam expert round Diss, and knowing so little myself, it had seemed prudent to seek him out. A tall lean man answered my knock, and when I had told him my needs, he did not hesitate. On the way to West Dereham that evening he talked of his past, or some of it, and how for years he had been a journeyman fitter for the now extinct firm of Youngs of Diss. If any engine owner within twelve or fifteen miles of Diss was in trouble, Jack was the man to put it right. His store of anecdotes was as vast as his knowledge and experience of engines, and the descriptive terms he used went far beyond my ken. In recent years he had been on both maintenance and construction work at a mill, riding the old bike he had used for racing forty years before. His agility both physical and mental were remarkable for a man of seventy. Full proof of that came later, but as he crawled over, under and inside those three rollers at West Dereham and pronounced them basically sound, I was happy enough to bank on his judgment and buy them.

When the first one had arrived, the front roller had to be taken off and sent back to West Dereham, so that the convertible tentonner could be loaded, since it was propped up on sleepers in place of its own front roller. By the time the next fortnightly opening Sunday came the three stood there side by side for all to see. All of them were very rusty, and though it wasn't difficult to picture them as fully restored and worth much more, it was much harder to convince others either that it could be done, or that it was worth doing. Of course it would take time and a great deal of effort to scrape down the rust and grime, to bring the deeply rust-pitted works back to shining steel, to re-lag, re-tube and to apply several coats of paint. But though I suspected what some of my family were thinking, and perhaps saying among themselves, they made a good show of being tolerant towards me whatever their private views. Flora had laughed when I told her I'd bought all three rollers. I fancied I detected amusement as well as apprehension in her tone. But Bridget, my eldest daughter home that weekend from London, had been rather scathing in her comments to Flora, later passed on to me. 'I rather liked the idea', she'd said, 'of Father being a bit eccentric in having an old engine for a pet, but to get those three heaps of junk is just crazy!' That's a pity, I thought, although having been pretty close to Bridget for the whole of her twenty-five years, I didn't think she would hold it against me for long. I meant to find out as quickly as possible just what the task of restoration entailed, tackling one at a time. Filling the boiler of the eight-tonner, I soon found that new tubes were essential. With Don Hubbard the lorry driver, and Ben Francis the pigman, also Percy Piper from the garden willing to lend a hand during the long mid-summer evenings, oil, elbow grease and brute force freed the pistons and rods, so that by turning the flywheel by hand, all the works could be set in motion. The twelve-tonner was then filled with water, and as there were no visible leaks the fire was lit and steam began to rise.

That evening a little blue Austin van drove in, and out of it stepped a slim man whose boyish-looking face belied his years. Approaching with obvious hesitancy he said quietly, 'I hope you don't mind, but I've had a bit of experience with steam and wondered if you might need any help—weekends or evenings? Roger Garnham's my name and I live at Garboldisham but work at Potter Bros. of Dereham, driving a steam roller most of my time.' Of course I didn't mind, I told him. He could not have turned up at a better moment because it was certain that with these three additions, I should at times need someone with the mechanical skill and knowledge I lacked. There was, I was fast realizing, a pretty wide gulf between enthusiasm and experience. Roger told me he had in fact once worked for Doran Brothers and knew these engines well. It was reassuring indeed to hear his opinion that although there were visible signs of decay, the most vital parts of firebox, boiler and motion should give no trouble. And if any re-tubing was needed, he would be quite willing to do it. In fact, he had quite a liking for the job even though most chaps didn't care for working in the cramped confinement of a firebox. 'One advantage of being little,' he said with a smile.

The twelve-tonner proved to be sound, tubes and all, though Roger had said they might not last long, having stood empty for years. It was a heavy ponderous engine to drive, but such machines were made that way, and I could see it doing good to some of our own roadways, now that we knew it could still work. But the convertible ten-tonner appeared to offer most scope, though it presented a problem. It looked so forlorn and incongruous standing there, propped up with oil drums and sleepers—an iron horse with no front legs. Knowing that a search for a proper forecarriage would be hopeless, I sought for, and finally acquired from Cecil Knights, a pair of iron wheels that would at least take the weight. They had come from an old threshing drum, and it took a lot of scheming to make them fit, as well as to attach the steerage chains.

With the engine movable at last, and some leaky tubes replaced with some sound second-hand ones from Mr. Palmer, I lit the fire one hot Saturday morning, determined to drive her that afternoon. After lunch, finding that with only thirty pounds pressure on the gauge I could scarcely expect to move her, I stoked up again, and went for a spell in the garden with a hoe. Passing the thermometer, standing at 89 degrees in the shade, I returned to find the engine registering 180 lb. pressure. This was disturbingly unexpected because the safety valve should have blown off at 160 lb. pressure. Here I was, I thought with growing concern, quite alone with an engine built in 1911, having made no other test for boiler and firebox soundness than tapping with a hammer, building up steam of a force that could quite conceivably mean an explosion. As I climbed up, the heat was almost overwhelming. I opened the firehole doors in the hope of seeing the fire going down, so that the danger would lessen before long. Only I could decide whether to try extracting some of the flaming oak logs through the narrow door, but the heat and my growing apprehension made me retreat. A few yards away stood the twelve-tonner, and from behind the protection of its heavy rear wheel, I could still see the pressure gauge. Obviously the safety valves had stuck, and should have been dismantled and checked before I raised steam for the first time in years. From this position I reckoned I'd be fairly safe if the boiler exploded, and I wanted to see just how high the pressure went. This would be a test indeed, I thought, but of a kind I would be certain never to repeat. Up went the needle to 190 lb. and then having just passed the 200 mark, half an hour later, slowly began to drop back. It took another half an hour to drop to the safety figure of 160 on the gauge; at this point, dripping with sweat and so thirsty I could scarcely spit if I'd wanted to, I began walking back to the house for a drink. When I met Flora on her way out, bringing me a glass of iced lemonade, it seemed senseless to recount the incident. After I had quenched my thirst I said, 'There's enough steam now, if you'd like to see how she goes.'

10

Beware of Ramps

It was rather humbling to realize how little I knew about its tricks and mysteries for all my love of steam, to say nothing of its technicalities. Though far simpler than the internal combustion or diesel engine in design, in practice steam engines can be much more temperamental. The two ingredients for generating power were just fire and water. But a fire could be controlled, with its only outlets for heat and smoke being a boiler, with its water-filled firebox cavity and its tubes leading through the barrel to the chimney. The draught could be regulated by the ash pan flap, though this advantage could be offset by sooty tubes. It was the means of supplying water against the pressure of the boiler that caused most anxieties. The Burrell Road Rollers I'd bought were fitted, like most traction engines, with both pump and injector.

Readers with steam experience will have already queried, if not condemned, the course of action I took when the ten-tonner failed to blow its safety valve. It could well be thought that instead of deserting the engine I should have injected or pumped in cold water from the tank, so as to bring down the pressure. This was, in fact, what I tried to do. The water level stood half way up the gauge glass, and with the safety valve stuck I reckoned it should maintain that level, though I could not be sure of this. But there was a gurgling noise in the water tank and the filler hole flap was warm and steamy. Only a slight leak past the appropriate valve perhaps, but it was enough to prevent the injector from working. Unlike the safety valve this had been dismantled, checked and replaced and would no doubt have worked had the water stayed cold. Of course the water valve too should have been checked to make sure it was seating properly, yet with such things only tests under pressure may reveal a fault. It was thus with the pump as well. This was driven direct from the crankshaft on an eccentric, so that the pump plunger worked whenever the flywheel was turning. Water was pumped into the boiler only when the bypass valve was shut off, but I was stumped when I had closed this valve and there was no sign of either air or water when I opened the pet-cock, placed there for testing and getting rid of air locks. This was the reason why I decided it wiser to quit. Someone more experienced might well have had no qualms and known better what to do in such a case, or he too may have judged it safe to leave, knowing that the water level could not become dangerously low. This in fact proved to be the case, but I had plenty of doubts at the time.

One tends with steam engines to learn the hard way, by practical experience gained in many varied situations. It may help to listen to old enginemen's advice, and yet so often one has forgotten it by the time the circumstances they describe in such fruity detail come about. I was learning, but was also being drawn towards the idea of building up a little collection. I'd been pretty lucky so far, and the discovery that engines were still to be had if one searched hard enough, at prices which must surely increase, was one of the spurs. The other spur, apart from personal indulgence, was the appeal they had to Sunday visitors. My latest acquisitions were of such a decrepit appearance that it was amusing to compare the quite different expressions on the faces of those who appraised them, with the admiration of those who gazed at Bertha in her shining green paint. I overheard such mutterings as 'Somebody'll have a job on to make this old thing look smart', and these came to my ears as a challenge.

Bertha was on show again next Open Sunday, and again I was enjoying giving rides to children and to an occasional adult as well, if I liked his looks. One such man came forward just after I'd decided there was no more room. I was about to tell him he'd have to wait till next time when he said he was not asking for a ride, but could I tell him where he could find the owner, and did I think he would be interested in a Marshall he wanted to sell? 'I'll be back in three minutes,' I said, blowing the whistle and setting off with rather a jolt. The boys behind had to hold on tightly, and it was a shorter circuit than usual this time.

The young man was still waiting, and with the boys off the tender, I invited him to join me. 'I thought you might be interested,' he explained, after I had told him who I was. 'You see, I want to get married soon, and I can't afford to keep a wife as well as an engine, and so my Marshall has to go. My visit today is pure chance. I have brought my future wife and her parents along to see your gardens, not knowing that you had engines too. My Marshall is a 1909 double-crank compound—quite a rare type—and I shall miss taking it to rallies, but there it is. The trouble is, I need the money pretty quick.' Trying again not to appear too keen, I asked a few more questions about its condition, location and of course the price. It surprised me to learn that he had come from near Lincoln, and had bought the engine only two years before; having repainted it he had taken it to rallies whenever he could. He was a railway signalman on a line that was to be closed. This and his forthcoming marriage meant a move to a new home, and hopes of being able to keep the engine had been finally dashed only a few days before. 'The only thing really wrong with her is her chimney,' he assured me, explaining that it had lacked one when he acquired it. He had found a second-hand one from some other make of engine and used it ever since.

It was then that he told me the price, adding that if I wanted to come and see it, would I do so within a day or two because he was in a hurry to sell. At that price no thought of trying to beat him down entered my head, for by comparison it made Bertha seem dear. 'Let's go across to the office,' I suggested. It was not far away and there I wrote out a simple agreement which covered us both, giving him a twenty per cent deposit and a promise of the balance by return on delivery. He was happy, and so was I, very happy indeed, at the thought of engine No. 5 of a different make and type, due to come as soon as I could arrange transport. I had taken some risk no doubt, but if on many past occasions I'd lost out when I had felt compelled to believe what a seller told me about something I could not prove, it could scarcely apply in this case. It was a gamble I could scarcely lose.

Yet when the Marshall arrived a few days later the very first sight of it jarred, just because it had been so atrociously painted. It was, to begin with, a harshly glaring green. Most Marshalls were painted a rich deep red, and I quite expected this one would be too, though the question of colour had not cropped up when we were doing the deal. What jarred even more was the way in which paint had been used to doll it up, as if by a very slapdash showman. Rivet heads and such like were of all colours, and it seemed so out of keeping for an agricultural engine that I wanted to repaint it at once. But since in the less accessible parts old dirt and rust had been smeared over with paint, it was certain that until a thorough scrape and clean-down of the metal could be undertaken, it would have to remain as rather an eyesore.

In spite of this first disappointment, the Marshall could be made an interesting and attractive addition, and though the first time we steamed her she made some queer thumping noises, Jack Clements said he thought he could cure them. I hoped he might cure her temperamental throttle too. Usually, a quarter of an inch on the throttle lever could make a considerable difference in the speed of an engine. The lever, called regulator as well as the throttle, which on some makes one pushed and on others pulled to stop or start, moves a rod passing over the motion to the steam chest. There the stop valve controls the steam passing through to the cylinders, and the throttle regulates the amount of steam which drives the piston, by opening or closing the ports. With this Marshall, you could pull back the regulator halfway, and nothing would happen. But another little tug of no more than a quarter of an inch would set the engine off at a furious rate. If in forward gear the lurch was so violent that the front wheels almost left the ground. I asked Jack Clements, who had come over to inspect it, what was the reason. 'I wouldn't say there's much wrong with the valve,' he said casually. 'I've known other Marshalls like that—it's just the slide tapers too sharp. What I'm more worried about is the crank-shaft bearings—something wrong there to make that clumpin' noise. I'll have a go at it if you like.'

Word that engines had begun to accumulate at Bressingham seemed to be getting around quite quickly. Several old enginemen, both drivers and owners, were coming to see them, and one of these was Leslie Aldrich, who had bought from Tim Stevens the Burrell which Leslie's father had purchased new in 1910. I found him a likeable man, as did others, and no one begrudged him carrying off several prizes which deservedly came his way at rallies. 'That's not a Marshall chimney—I expect you know,' he said as soon as he saw the new arrival. When I replied that this was one of the faults I would have to try and remedy, he followed it up by saying he had a Marshall chimney. 'It stands in the garden, with one of the wife's geraniums growing out of the top,' he added with a very broad smile. 'I daresay I could find something else to put there instead next year—if you'd like to have it, that is.' I certainly would, and when later the geraniums were taken out, Leslie brought the chimney over and refused to take a penny for it.

Well before that, engine No. 6 had arrived and I'd had to take out a bed of nursery plants to use the space for engines to stand on. It was an ideal spot, adjoining the standing I had made for Bertha, and still beside the path I used so often between the house and the nursery office. The Rev. Mr. Stebbing had called twice after our meeting at Woodton Rally. Early in August he came a third time, not only to see the Marshall but also to give me a piece of news he'd just come by. 'You remember me telling you about Bessie, the sister to my ploughing engine Cissie?' I nodded, wondering what was coming. 'Well,' he went on in his unparsonic voice, 'I hear that Duce's scrapyard has got to be cleared out before long, which means that if he doesn't sell the engines he's got there pretty soon he'll cut them up. He's got two or three others besides Bessie, but I'd say she's the best of them. You seem to be building up a collection, and I thought it would be rather nice if the pair could be saved, and be neighbours at least.'

So did I. It was true that since the Marshall had been almost thrown at me, the idea of building up a collection had been growing like a maggot in my mind. I'd scarcely given Bessie another thought, but now it suddenly became far more than just a thought. What I had rather drifted into as a hobby, began to crystallize as a worthwhile ambition in itself. Bessie was, I knew, an AA model Fowler, the largest type of ploughing engine made for use in England. They were massive and very imposing, and to have one would enhance tremendously the nucleus of the collection already here.

Although I knew Cambridge so well, since coming to Norfolk I had seldom been there, and until recently never knew that this yard tucked away between the railway and Coldhams Lane had been a crematorium for a good many engines in recent years. Richard Duce had been only a name to me, dating back from before the war, but since then he had gone in for scrap on a very large scale. When Mr. Stebbing and I called there next day, it was hard to find a place to park the car in between great mounds of scrap of all kinds. As an expert on Fowler ploughing engines, Mr. Stebbing had been there before, and having been given leave by Mr. Duce, we began exploring. We saw three other ploughing engines as well as a very large Portable. 'There was a time, I gather,' the Rector told me, 'when you could have had your pick from thirty or forty engines in this yard. Nobody wanted them, so they were cut up. Now there's only these three and Bessie which stands behind that big pile over there.'

And there she stood, complete except for her underslung winch drum and there were even the remnants of a tarpaulin over her top works. 'Came from Baldocks of Haverhill—along with three other pairs.' Mr. Stebbing was very well informed, and explained that when this well-known ploughing contractor's equipment was auctioned six or seven years before, Mr. Duce had bought practically the lot. Cissie was parted from Bessie soon afterwards, having been bought by Andrew Fountaine of Narborough for work in a chalk pit. Mr. Fountaine had sold it less than a year ago to a Norfolk scrap merchant, and on learning this Mr. Stebbing had offered to drive it from Narborough to Wymondham. But it never reached the scrapyard there because on the way he came to terms with its new owner, and drove it on to his own parish of Tacolneston—a week or two before the Woodton Rally, where I'd seen it in steam.

There was no beating about the bush with Mr. Duce. The price he asked for Bessie was reasonable enough, and so far as we could see it was in fair condition in spite of having stood there for so long that the wheels had sunk in somewhat under its twenty-one tons weight of metal. The element of a gamble was there, but with enthusiasm still rising and with my bank account recently augmented by book royalties, I was willing enough to clinch the deal, even though I failed to get Mr. Duce to budge. With a dealer one feels obliged, as a matter of course, to offer a figure lower than the price asked. It had already occurred to me that transport to Bressingham would be pricey, so to begin with I tried to bargain at his own price, delivered; but I saw from the look he gave me that this was a mistake on my part. 'Nothing doing,' he said bluntly, adding that it would cost quite a bit for him to get such an engine pulled clear of the great piles of scrap which hemmed it in, so that it could be loaded, and I must make my own arrangements after that. I had to climb down, but was given the promise that the missing winch would be replaced.

I also had to make several telephone calls before finding a haulier with a low loader of sufficient capacity. Finally two quotations came up, one for £36, the other for £18. Here I found, as

I have several times since, that some interest in steam preservation can make a vast difference to the cost of services. One firm had none, the other had already bought an engine for preservation and was willing to help someone else with similar objectives by charging at cost.

A month elapsed before Bessie arrived, and I fancy it would have taken even longer had not Mr. Duce grown tired of my importunings to release her from the confinement caused by being so hemmed in. When family inspections had been made, as she stood there majestically—twice the size of Bertha—Jack Clements and I began helping the driver to unload. With the low loader's rear wheels aside, and the platform eased down by jacks, the engine had to be held to the now sloping platform by the lorry winch and chocks. Ramps for the wheels were in position, and with me at the steering wheel of the engine, it was my job to guide it down as the winch was slowly released. At first it was lowered inch by inch, but the lorry driver turning the winch made a mistake as the six-foot rear wheels gained the ramp. Just at this trickiest moment the ratchet of the winch was released, and Bessie descended the rest of the way with such a rush that the steering wheel wrenched itself from my grasp. The front wheels, askew, left the lorry without touching the ramps at all, and slumped violently on to the ground. 'Cor, that was a near thing,' growled Jack Clements, under his breath. 'He never should have let go like that, how you kept your balance stuck up there I don't know.' I laughed, saying 'All's well that ends well', for in fact I'd not had time to sense the danger. Everything had happened much too quickly and unexpectedly for that. No damage had been done. With the lorry gone, Jack and I crawled over Bessie's grimy body. The pistons were seized up, but when he had examined it thoroughly Jack said he saw no reason why, once these were freed, she should not be tried out in steam.

It was engine No. 7 that helped to prepare Bessie for her first trial in steam for seven years. Within a day or two of this momentous arrival a portable engine had been offered, when its owner paid us a visit. Jack and I went to look it over, and once again I feel, simply because it was there, and representative of a very different type of engine. It was not self-propelled and therefore had to be hauled, in its day, by a team of horses for threshing or saw bench work. Partly because of the strength needed to move its six-ton weight over rough or muddy ground, it had neither coal bunker nor water tank. These essential steam-raising elements had to be brought to it, but at least it had its own pump for the boiler. It was another Burrell. Jack said that Burrell's made very few portables and this one, being a single-crank compound, must be rare. His assessment later proved correct, because it was discovered to be the only Burrell portable left in the country. I also found out that it had never gone more than a few miles from Thetford, where it was built in 1900, and our tractor driver who towed it the twelve miles to Bressingham said he could well understand why. The deal made no deep inroad into my resources, since Portables were of no interest to those who wished to drive engines at rallies.

Just at that time we were having much difficulty in freeing Bessie's seized-up works. Penetrating oil had been squirted into every movable part, and most of all in the cylinders, but still she refused to budge, until one evening four of us used a long pole as a lever through the flywheel spokes and wriggled it for all we were worth. At last the pistons moved a fraction. Continuing the struggle for an hour we got them free, but still so stiff that the idea occurred of using the Portable, with a belt from its big flywheel to Bessie's, hoping it could be freed sufficiently to make steaming safe. It worked, though the connecting belt slipped badly at first, and gradually Bessie's moving parts responded, helped by copious doses of oil squirted wherever it could be taken in.

All this may have been a needless precaution. Jack Clements had become more often available as adviser, and I felt that whatever opinion he expressed as to the right thing to do, we would be safe in doing just that. He reckoned Bessie's 'top works', as he termed all the moving parts assembled above the boiler, to be sound, but if we forced movement too quickly without the full benefit of oil so long absent, some damage might be done. With a modest head of steam Bessie's throttle was opened gently, and without a sound, with scarcely a wisp of escaping steam, the twin-cylindered engine, with its fascinating array of cranks and valve rods, began working. It was wonderful to behold, and almost impossible to believe, that all should work so silently and smoothly, with the throttle control so perfect. Though there came a time when Bessie nearly drove me to distraction, as I struggled to drive her, for the present I was delighted and lost in admiration.

11

A Frustrating Foden

For the whole of my life I have suffered from excessive enthusiasm when following up what I felt was a worthwhile notion or project. It might be termed a characteristic flaw or weakness to allow oneself to be so taken up that one can think of little else while the craze lasts. Some which I can recall did not last long because they proved impracticable or after all not worth the effort. Sometimes, too, the scope of a particular objective was limited, and as soon as I had achieved it, my mind would rest again until something else cropped up. The more mature I became, through lessons learned the hard way (such as the Canadian venture) I was able to inject rather more common sense before embarking on something that might end in failure. But the trouble then was that if, having weighed up pros and cons more thoroughly, I decided to go ahead, any unexpected obstacles became all the more of a challenge to be overcome.

With steam engines, however, it was more the development of an idea which snowballed along, and each acquisition gave me not only the taste but the excuse for more. At the beginning of 1961, I was fired only with the thought of having one traction engine. Well before the end of the year I had seven, and was losing sleep because No. 8, which I was trying hard to acquire, was proving so elusive. It was not a question of price. My resources had of course slumped to a pretty low level, and might involve going some way into the red at the bank, but the real anxiety came because the owners would not name a price. A visitor had given me the news that in the yard of a large works-contracting firm, near Norwich, stood two Fodens. They had been used as tar tankers for road making, and had been discarded only a couple of years before. My informant had worked there, and believed both had been bought new as steam lorries thirty years or so ago, by the same firm.

Visions from the past of Foden steam wagons came up to taunt me, and I remembered how, as a boy, I used to look out every Wednesday for Russell's Foden delivery wagon coming to the village. And in the 1920's when I was on long cycle trips there was sometimes the joy of hanging on to a Foden's tailboard for mile after mile. Their speed was just right at about 12 or 15 m.p.h., and if one of them passed me, it was worth a spurt to catch up with it and hang on as long as possible, steering the bike with one hand, and with the other holding on to the lorry. There were no restrictions on such practices in those days and as well as the pleasure of being pulled along, there was the effect of a sniff from the smoke wafting overhead, giving me the feeling of being steam hauled.

Fodens of Sandbach saw the red light in the early 1930's when steam engines for road haulage became so heavily penalized, especially on solid tyres. When I went to see those near Norwich by appointment, I gathered that they had been retained in use for so long because as tar sprayers steam engines could be used both for propulsion and for heating the tar. But now they were a sorry sight, caked with tar where not rusted. Flora had come too, and having made a joke about them in relation to my enthusiasm she wished me luck, as off she set, pushing Anthea in her folding pram for some shopping in Norwich, where I'd arranged to meet her later. This was because the firm's director responsible for the sale of redundant equipment was out. I was told he should be back in about an hour, and had decided to wait for him. But he didn't come, and with Flora and Anthea I went on to Surlingham where Mr. and Mrs. Goodings, the B.R. divisional manager and his wife, were expecting us for tea.

With tea and a look round the garden over, I telephoned from there and was able to speak to the director I waited to see. As I had been on edge for three hours already, in hopeful anticipation, it put me even more on edge to be rather curtly told that it was up to me to make an offer for one or both the Fodens. I had expected he would simply name a price so that I could at least know whether or not I could have it. This was disconcerting, and I hesitated. 'Well?' came the voice, 'I expect you know what you're prepared to pay, don't you?' Knowing I had barely £200 left to spend, I had to decide quickly, and pointing out that as it stood, chimney-less, with its cab wrecked, I'd offer £100 provisionally for one of them. 'Very well—I'll see whether the other director concerned is agreeable, and will let you know.' With that he rang off.

The next few days were difficult. I could not dismiss the Fodens from mind and wait patiently. After two days of frustrating silence, and one night largely spent in worrying out a means of obtaining an answer, I decided to telephone again. The first four calls, made at intervals, failed to locate him. The fifth reached him at last, and there again was the same testy voice. 'Oh! that!' he said as if it were of no consequence at all to him. 'You wanted to buy one, if I remember rightly—well there's nothing doing.' With my hackles rising I asked why not. 'You can't have one, that's all. They were both sold to Mr. Cushing two years ago, which reminds me I must ask him to take them away.' I had nothing left to say and hung up, wishing I could have blown my top yet knowing that would do no good at all. It left me with a feeling of utter frustration, at least for a time, till calmer views began to prevail.

The name George Cushing was already almost a byword in Norfolk. It had cropped up so often since I began my quest for steam, because he was reckoned to have by far the largest number of engines in the county, if not in England. Some, including Thurston's famous Burrell Showman's 'Victory', had been beautifully restored, it was said, but many (including others of Thurston's fleet) were standing outside and deteriorating. It was also said that George Cushing could never bear to let an engine for sale slip through his fingers, nor could he be persuaded to part with one he'd already bought. But engines were not his only love, and more recently he'd gone in for fair and street organs as well. These he played on Sunday afternoons and allowed people to listen to them.

It was Mr. Cushing's reputation for not parting with anything he had acquired that brought on the dismay I felt on learning that the Fodens were his. I had never met him, but on calming down I knew that the only course left was somehow to make contact and hope he could be persuaded to let me have one of the pair. A letter, I thought, would be best, knowing that when words have to be carefully chosen I need time in which to choose them. In this letter, I explained how I'd learned that these two Fodens turned out to be his, and that I'd set my heart on obtaining one of this make, which would go a very long way towards making my little collection complete, though I had no wish to compete with him as a collector.

Flora read it before it was sealed. She laughed. 'Maybe he'll believe that bit about making the collection complete, but not me. That's what you said when you bought Bessie and I'm beginning to wonder if it will ever be complete. Don't forget about relaying the kitchen floor, and the spare bedroom is due for redecorating, and all the house windows need painting.'

Such reminders had to come, for they were justified enough. One of the penalties of a big house was the upkeep, and with Flora not one to nag, I must watch my step. But just this one more, I told her, and then I'll be content if for no other reason than not wanting to tackle more than we could manage to restore during the winter evenings. I could still reel off gardening articles to bring in money, though these too were for the winter evenings then fast approaching. The thing was that steam engine prices were still low, and I was certain they would rise.

Next morning, the telephone rang while I was having breakfast. 'I got your letter,' came an unfamiliar voice, 'and when I get those Fodens here, you can come and pick which one you want. I can't tell you the price just now—it depends what they cost to bring them to Thursford, but I won't hurt you over that.' The niggly suffering imposed by the urges from within which I seemed so powerless to control had lasted over a week. And now suddenly I was lighthearted again. I'd begun building another flint wall in the garden, a job that often had the same diverting effect for me as a jigsaw puzzle has for others. Like jigsaw pieces, the odd-shaped flints sometimes proved slow and difficult to place, but that day every stone seemed to fit.

We had often spoken of going to Thursford to see George Cushing's engines and to hear his organs. The message from him a fortnight later to say that the Fodens were there gave an added purpose to the visit we made that Sunday afternoon in October. Jack Clements came too, for on reaching seventy he had decided to retire from his job at the Mill and come on a short-week basis, to lend his much-needed skill as a steam fitter at Bressingham. The journey of fifty miles across Norfolk from south to north ended at Laurel Farm, where already a number of other cars were parked. The sound of organ music was thrilling and the sight of the huge and very ornate instruments was most impressive. Flora especially was enraptured, but Jack and I were very taken with the splendidly restored engines—seven or eight of them standing majestically in the one large shed with the organs. One could dismiss the sight we'd had of a score or more engines left outside, and apparently spoiling, when those inside had been so painstakingly restored. 'What he does,' Jack remarked, 'he does well and no mistake.'

There was no mistaking Mr. Cushing either, since it could only be he who was plying the big Marenghi with card music. He told us where to look for the Fodens, and there Jack was once more in his element as he carefully examined each one in turn. His verdict was that the one lacking a chimney, and with the cab stove in, seemed to be in better condition as far as works and firebox were concerned. Later, with his popular organ recital finished for another Sunday, I put the inevitable question to the quietly spoken George Cushing as he was fastening the door of his big shed. 'You can have your pick for a hundred and sixty,' he said, and at that out came my cheque book once more.

Taking the cheque, he gave an amused chuckle. 'I'm glad you're having it,' he said, 'because there was a man here yesterday who got a bit shirty when I told him I'd promised you first refusal. I won't mention any names, but you may guess when I say he's done a bit of dealing in engines for several years, but he makes a good job of those he does up to sell again. He said he couldn't understand me letting a flower grower who knew nothing about engines have one. He didn't think much to that at all.' I was chuckling too, to myself, but reassured him that I intended making a good job of restoration myself.

'I believe you will,' he said, 'and it'll be seen by a lot of people who come to your garden. I may have a go at the other Foden myself—but if I don't, and decide to sell that too, then it won't be to him.' No name was mentioned, but I knew George Cushing was referring to the same man who tried to sell me a Garrett long before I bought Bertha at half the price he was asking then. George Cushing was a man much more to my liking, and I felt then that we had a good deal in common as far as our approach and objectives were concerned, though he had been more shrewd and longsighted than I. I learned that he too had known what it was to struggle, having begun on his own account with one old steam roller which he hired out on road work with himself as driver. I discovered also that this linked him with the big contracting firm which had owned the Fodens, and it gave the clue as to why he was able to speak for them so long in advance of me.

My luck had held, and I could have no complaints even if

problems of my own making were piling up. One of these was the lack of a chimney. A steam engine without a chimney is totally incongruous and it must be the correct style for the type. The brass or copper top to a chimney made all the difference, yet though the chances of finding one for the Foden were remote, so was that of finding a craftsman who could make one, even though such a person might exist somewhere. This applied to a large extent to spare parts for engines. It was a matter of luck if a need could be met, and though Jack Clements was clever enough at making certain parts, we lacked the machine tools on which to make them. It was still possible to have castings made, and there was indeed a foundry in Diss where, so long as they had a pattern, iron could be cast into such things as firebars and hub-caps. The best chance still lay in finding from stores bits and pieces not yet sold for scrap, and one such store stood at Thetford, in the yard of Doran Brothers, who had owned the three Burrell Rollers I'd bought. At one time they had had a fleet of ten or twelve rollers, and the contents of the stores had remained after Mr. Palmer had bought most of the engines for scrap.

We made an appointment with Mr. Brown, the now retired partner in the closed-down firm, and Jack and I were able to browse through the contents of a long wooden shed. It was chock-a-block still with a vast amount of fittings, but when it came to picking out what we needed against likely future requirements, it was rather disappointing. Some suction hoses, fusible plugs, various pipe fittings, tube expanders and the like were placed in a heap, but we found little of which Jack could gleefully say, 'Now that's just what we want.' The purchase came to only three or four pounds, and having paid him I walked over to two Burrell Rollers still standing in the spacious yard now rented by a firm of coach owners. 'They're the only two we kept back,' Mr. Brown explained sadly, 'my partner and I let the others go to Mr. Palmer, and the price he gave was so low that we decided to keep one each for old times' sake; but it doesn't look as if we shall ever run them again.' He knew, of course, that three of his former fleet of rollers had come to Bressingham, and this had pleased him. Impulsively I asked him to let me know if ever he decided to sell. 'It's hard to tell,' he went on, 'it's a matter of sentiment more than anything. I've always said no when asked if I'd sell. It's nice to see them here, but you shall have first offer if we decide to let them go.' I thanked him, not expecting such a promise, and not hoping that the offer would ever materialize. We stowed the purchases in the boot of the car, with rain beginning to fall quite heavily. 'I'd better take shelter in the big shed,' he said, indicating a much larger open-sided building that we had not entered; and then he added, 'I believe there's a Foden chimney in there—I don't suppose that would be of any use to you, would it?'

12
A Slow Journey

One of the few features I had not liked at Bressingham was the closeness of the farm buildings to the house. A small backyard was hemmed in by buildings which had been the usual offices of coach house, stables and stick shed. The last had a loft, for it was part of a much higher barn, and as all these sheds next the house were narrow, the purely farm buildings which faced the other way were not in fact many yards from the rear of the house. On warm summer days flies and smells wafted in, yet with the area of buildings inadequate for the size of the farm, animals had to be housed there until capital allowed new buildings to be erected farther away. The existing ones were old buildings, with flint or red brick walls and pantile roofs, save for the big barn which was roofed with slate. This must have been there for two centuries, and for all its height was much less usable now than in bygone days when grain was threshed out on the floor by men using flails. It had a pair of high, wide doors through which carts loaded with sheaves could pass, and through these doors, flanked by cattle on one side and by pigs on the other, the derelict Foden was pushed soon after it was towed in from Thursford that November. It was the only roof under which restoration work could begin, and until an electrician could be found to fix some lights we used candles to light our labours. Daytime work was out, with routine jobs on nursery, garden and farm more essential. In any case the barn had no windows, and daylight could only come in when the big doors were open.

It had already come as a very pleasant surprise to find that some of the regular staff had volunteered to help on restoration work. Percy Piper, and young Dennis Leeper from the garden, Don Hubbard the lorry driver, and Ben Francis the pigman had helped on various jobs of their own accord while the evenings were light. Now winter had come they were still willing, despite the dim and draughty barn and despite my decision to tackle first the Foden, the most formidable restoration task. For a couple of hours on most weekday evenings we scraped off caked greasy dirt and chipped tar-bound wheels and frame, dismantling all except the engine, which had to be left to Jack Clements's expertise. At half past seven I used to leave them, returning a few minutes later with beer or coffee—a service Flora had rendered until babies took priority. This, and the satisfaction they seemed to gain from such tedious work, was their only reward. Even if later on one or two lost their first enthusiasm, I could not blame them, any more than I could feel ungrateful for what they had done.

Neither could I blame my two sons for not sharing my enthusiasm. Both returned home as 1961 came to an end, Robert the farmer and Adrian the nurseryman, though the latter had one more spell abroad the following summer before settling down for good. For years I had feared they might choose some calling which would leave the future of the farm and nursery blank beyond my lifetime. Now their minds were made up in just the way I would have wished, and whatever their private opinions about the place my steam engines should occupy, they were at least tolerant towards my liking for them.

By the time the Foden, with cab and chimney restored, was ready for a trial in steam; it stood proudly with every spot of tar replaced with bright red paint. It still lacked a body, but there was one waiting outside ready to be lifted and bolted on. Finding this had been another stroke of luck, for though it was never intended for a Foden, the ex-railway lorry body, with scarcely a fault, was an exact fit. Roger Garnham was no stranger to Fodens as a driver, and it was he who drove it first one Saturday morning. The cramped little engine, topped by a multiplicity of working parts, was rather too complicated for me to drive without some practice, but Roger knew how to handle Boadicea, which was the name we'd given her. My chance came that afternoon, when no one but Robert and I were there, and we were amazed that so small an engine could be so lively. Jack had done his work well, making good faulty joints and bearings as he stripped down and re-assembled almost every moving part.

The first fortnight of March 1962 was cold. Sharp frosts every night followed clear days, and though the damage this spell did to the nursery business was felt for the whole year, it came as a gain to the restoration programme. Bessie had been steamed into the big barn as soon as Boadicea was out. She had to go in backwards because the barn was so narrow, and her massive six-foot-high wheels would have blocked the doorway. Besides, these wheels had to be removed, and handling them in such confinement to get them clear of the jacked-up engine was a tricky and sometimes dangerous job. Her weight was too much for the thin asphalt floors of the barn, and to sort out such snags took up precious time. And we were racing against time, for once the frosty nights gave way to normal March weather we would be up to our necks in normal early spring work in garden and nursery. There was so much of Bessie. The top works were good, we knew, and only needed cleaning free of rust. But for the rest of her, tank, wheels and the huge underslung winch drum, it was a case of scrape and rub, inch by inch, till clean enough for painting.

We decided Bessie should be blue. She had originally been black, but we learned from Mr. Stebbing that Fowlers turned out engines in any colour specified by the customer. We could exercise the same prerogative now. From grey primer to top coat of royal blue, we went over Bessie four times with the paint brush, and though spring did not come until after the cleaning down was finished, I took time off on Good Friday to complete the last coat of paint so as to have her ready to steam, as a special treat, on Easter Monday. Bridget was home that Easter weekend. With Bessie in steam both she and Flora came to inspect, now that the engine was out of the murky barn and in the sunlight. Flora had one-year-old Anthea in her arms, and all three joined me on the footplate. It was pleasing to hear their admiring comments but Anthea stared in silent wonderment at the moving parts. 'Perhaps you're not quite so crazy after all,' said Bridget with a smile, adding a request to be taken for a ride.

Spring, as always, made its well-nigh total demand on the time available. For a nurseryman with a five-acre garden to look after as well, April and May were crucial months. But with the approach of Whitsun I began to look forward more and more to the adventure I felt sure it would be to drive Bessie to the annual Woodton Rally on Whit Monday. Mr. Stebbing had volunteered to drive, since he had Cissie half stripped for restoration, so Bessie would be the only ploughing engine there. She looked smart enough we thought, to gain the prize for the best engine in the parade, but we must allow ample time for the journey of twenty-six miles.

These big Fowlers are not the easiest engines to steer, and as I was of much more robust physique than the Rector, I took my stance on the steering wheel, with legs astride the tender for balance. There was no question of sitting down and it was soon evident that muscle, alertness and judgment would all be needed. It was at nine o'clock on the Whit Saturday that we edged out on to the main road. Despite holiday traffic on roads leading eastwards to the coast, the police had advised me not to take the risks attendant on taking such a wide vehicle on to by-roads. The first of my worries were 'cats' eyes'. The engine was too wide to keep between them and the curb or grass verge. In places I could not prevent the off-side wheels running over them, since to straddle them gave oncoming traffic too little room to pass. If I tried to keep Bessie close enough to the verge to avoid them, the near-side front wheel was just as likely to mount the verge, and if it did so no power that I could exert on the steering wheel would guide her back on to the road again. She would simply draw farther still off the road, and only by stopping and reversing could I bring her back. Maybe the steering chains were a bit slack. Maybe I lacked experience as a steersman, but since the slightest alteration in direction meant turning the handle two or three times, it was small wonder that to keep exactly to the foot or so next the verge was almost impossible.

Beyond Victoria Road railway bridge, Jack Clements was waiting for us, as he had said he would be. It had taken nearly an hour to travel that three miles in spite of the fact that flat out, Bessie could do ten or twelve miles an hour in top gear. The Rector had already reported that one of the two injectors would not work. Jack said that was one thing he had not felt very happy about, and because of his general anxiety he had decided to come too. 'Wait a minute while I get my old bike,' he asked, running smartly down the path to his nearby house. He was back, wheeling his old racing bike, before I could explain that he had no need to bring it. Flora had already promised to come if she could, or Don if she could not, and bring the car to Woodton in response to a phone call to say that we had arrived. 'No,' said Jack, 'I'll ride me 'ole bike—it'll be handier if there's any trouble.'

For the next few miles, crossing the main road at Scole and taking the east coast route to Harleston, there was no trouble, at least not as far as the engine itself was concerned. Jack on his bike was rather like a shepherd dog, sometimes in front, sometimes beside Bessie, but always on the alert for trouble. Every now and then we had to stop to back off the verge which I'd allowed Bessie to mount in my anxiety to allow cars enough room to pass. Sometimes when a queue of cars was building up behind we pulled in to let them through. I'd stopped worrying about smashing 'cats' eyes', after checking that the first few we'd run over appeared undamaged by Bessie's great iron wheels.

'We ought to stop for water at Harleston,' the Rector shouted. You had to shout to be heard above the clatter. The first big garage in Harleston was owned by Cecil Knights, the Norfolk Steam Engine Club's secretary, and his brother, and it seemed the obvious place to stop to fill up Bessie's tank from the supply they were sure to have. But Cecil was not there, having gone to Woodton with his own Garrett engine. His brother came out and said it would be better if we took a side road where water could be had without hindrance to the traffic now thickening fast, since the tank might take half an hour to fill. 'But it means backing and turning round to get there,' he said, and he added a warning to avoid running over the iron inspection lids over the underground petrol tanks supplying the range of pumps. 'How's the water?' Jack Clements called up to the Rector. 'Pretty low,' was the Rector's reply, as he pulled back the reversing lever, while I slewed round the front wheels.

The road was narrow, and the slope up which we had to back was steep. The traffic was bothersome, and one shunt was not enough, with these petrol tank covers to be avoided. These factors, as well as the noise, made co-ordination between driver and steersman difficult, and when Bessie ran down forwards from the slope for the second shunt, it was too fast for me to get the needful turns on the steering wheel. It was too fast for the narrow space as well, and one front wheel caught the opposite curb with a thwack which made Bessie shudder. It made me shudder too, when I found the steering jammed as a result. 'Back right up this time,' I shouted. We were blocking the road completely at right-angles, and nothing short of backing into the yard of the garage could save the situation, since the steering must be freed before anything else. Already I was feeling aggrieved because all this trouble would have been saved had we not been asked to turn to go elsewhere for water.

In this present position I noticed that we would be well clear of the petrol tank lids, and once up the slope and on the level again, the tank could be filled just the same. But our troubles here were not ended. They had in fact only just begun because as Bessie's front wheels left the road, allowing the piled-up traffic to move once more, there came a sudden roar and a gush of steam through the firehole door, which shrouded Bessie instantly. The Rector and I left our places rather hurriedly, but Jack was the first man I met through the cloud of steam. 'Fusible plug,' he announced bluntly. 'Stands to reason—short of water, back up the slope—head down, no water over the crown of the firebox—fusible plug blows—simple as that. That's what fusible plugs are for, but it shouldn't have happened none the more for that.'

The Rector had little to say. As driver his was the onus, and perhaps the shame, of having dropped a plug, since for old enginemen this occurrence is a reflection upon skill, and under no avoidable circumstances should a driver take a risk when the water level reading is low. But these circumstances were unexpectedly difficult, and it was not for me to blame him. Nor could I blame him for allowing Bessie to run forward so fast as to hit the opposite curb, because there was excuse enough for getting flustered with so much fuss and bother.

With the lead-filled core of the fusible plug melted, the hole allowed all the steam in the boiler to squirt downwards on to the fire. The fire was quenched long before all that steam had spent itself, but by the time we ourselves had swallowed some much needed drink, Jack had gone to work on the steering. The worm gear was jammed, but in half an hour Jack had freed it. 'Let's have a bite to eat now,' he suggested. 'We've got to wait till she cools down enough for me to get that fusible plug out.'

It was of course the only way to get on the move again. The plug must be unscrewed, and refilled with molten lead, and not until it was replaced could the boiler be filled again, and the fire re-lit. I had reckoned on being the one to crawl under the engine to remove the heavy ashpan and firebars, so as to get at the plug, since there was no other way of doing it. No one else could be expected to do what was bound to be a very sweaty, dirty job. But I hadn't reckoned with Jack either, for though he had turned seventy he could be very determined, and he had made up his mind it was his job. After we had eaten our lunch sandwiches, Jack said it would be cool enough to begin. The ashpan was dropped and pulled clear; one by one the still hot firebars were hauled out, each one making us splutter with ash dust. An hour later, the plug was unscrewed. One of the garage men melted some lead in a crucible and filled up the hole. By three o'clock the fire was lit once more, and by half past four we were again on our way to Woodton, with Jack back on his bike as shepherd. The injectors were still giving trouble, and several stops were made while Jack tapped and fiddled with them. They seemed to take it in turns to fail, but what we dreaded was that both should refuse to work. Beyond Homersfield we stopped at the 'Dove' Inn for tea, and were allowed to drain a handy water butt to replenish the tanks.

Flora arrived in the car, saying she had found it easy enough to follow our trail of wheel and water marks. On again, we were soon able to turn off the still busy main road, and headed towards Earsham. Bessie's speed had increased, since now the road was mostly ours. But hopes of reaching Woodton that day faded. With four miles still to go, I was given permission to pull into a farm yard as dusk was falling. With Flora and the car still in attendance it was the sensible thing to do, and though we did not care to admit it, Jack, the Rector and I had had enough for one day.

It was not until early on Whit Monday morning that Bessie finished the journey. Once on the big meadow at Woodton Old Hall, with thirty other engines already there, we began furiously to clean off the dust which by now had dimmed Bessie's shining paint. The judge came round. 'What a pity you painted her blue. Fowlers should always be either black or brown—didn't you know that?' was the muttered comment. When the prize winners were announced, Bessie was not among them, though it was whispered that but for the colour, she would have gained the rosette for the smartest engine there.

Next day, with Jack driving and me again at the wheel, we took the byroads back to Bressingham. Jack did not reckon to be an engine driver, but if he saw a pond or a ditch, or any sign of water, he'd stop and say, 'Let's top the tank up—there's no telling when we shall find any more. I don't want that plug to drop with me. I'd never live that down.'

13
Penny wise—

Regular Sunday openings had to come and by 1962 they began on the first Sunday in June. During the latter part of the summer of 1961 it had been almost as much trouble on the alternating closed Sundays to keep people out, as it would have been to let them come in. Besides, I could sympathize with those who had travelled some distance only to find they had chosen the wrong day, for though openings were advertised, not everyone saw the East Anglian papers. The number of visitors was undoubtedly increasing, and with eight engines all told, there was an excuse to steam one or two each week in turn. As the afternoon crowd built up there was now less of the car park free for giving rides to children, but the demand was sometimes embarrassing, and though the garden was still the greater attraction the interest shown in the engines was enough to set off my collecting urge once more. Not that it had ever left me, despite what I'd said about the Foden completing the collection. I knew now that in the full flush of my yearning to have it, this had appeared at the time as the apex of ambition. Once I had acquired and restored it, the void had reappeared, with the knowledge that several other types of steam engine were not yet represented. On these grounds the collection was by no means complete.

Last year, I remembered pooh-poohing the suggestion made by a visitor that I ought to have a Showman's engine. I told him these did not really appeal to me, since they were so ornate, and not linked to agriculture. But this, I saw more clearly now, was a case of 'sour grapes' because I knew Showman's engines were more costly than any to buy, and would be quite out of reach. There was, too, some lack of confidence in our ability to restore a 'Showman's, even if one derelict at a low price could be found. Now, having made a fair job of the Foden especially, it was clear that we were capable of doing the restoration. What was more, it appealed far more to me to buy an engine in need of restoration than one in good condition, even if I could have raised the much greater cost the latter would entail.

'I wonder if George Cushing would sell one of his,' Flora said. I'd told her what was biting me, and had been able to think of no one else who might have an unrestored Showman's for sale. Apart from his splendidly restored 'Victory' George Cushing had three which he had not been able to tackle since buying all Thurston's fair engines several years ago. 'You can but try,' Flora had suggested, gently reminding me that my collection was supposed to be complete. But when we visited Thursford again a few days later, George was reluctant to commit himself. He had hopes, he said, of restoring all three: Unity, King Edward and Alexandra. But he'd think it over. He showed us round his organs again and played a tune or two especially for us that Saturday afternoon. He took us to some delightful woods he'd bought, explaining that he'd always wanted to own such a place where trees and flowers grew as nature intended. Beyond that was a little yard, well hidden from the road, and inside were other engines some of which he had acquired years before. 'That Clayton wagon over there,' he said ruefully, 'I used to drive myself thirty years ago to London. Those were the hard times, and it was a job to make a living. There it stands with a tree growing through it, beyond repair I reckon. I never could bear to part with it, but I'd have taken more care of it, and the others, if I'd ever thought there would have been such interest in them again. It was like that with Thurston's Showman's engines, I bought them just because I couldn't bear to see them go for scrap, so you can understand how I feel about parting with an engine. I've hung on to steam for love, not for money.' Unity was one of Thurston's engines there, and as I cast longing eyes over her rusting body he said, 'I'll tell you what I'll do. If you can't find one in the next few weeks, I think I could let you have one of mine. You advertise for one, and if you want me to come with you to look at one, you've only to let me know.'

This was fair enough. It wasn't the blank refusal I'd been told George Cushing gave to would-be purchasers. A few weeks later I received an offer from a Mr. Evans of a Burrell Showman's named Black Prince, described as being rather rough but capable of being restored. It was an engine with a past, for since 1904 it had been used for haulage and later converted to a Showman's for Grays of Hampstead Heath fame. During the war it had been used for pulling down dangerous buildings following air raids on London; since then it had lain discarded in breakers' yards at Twyford and Ewell. Mr. Evans was a scrap merchant, but it seemed he was alive to the reawakened interest in steam.

'I'll be there at ten o'clock in the morning,' George Cushing had told me over the telephone when I reminded him of his promise. Though it was noon before he arrived, we spent a few interesting hours in Mr. Evans's yard. Black Prince was hemmed in, but at least it was under cover. George knew what faults to look for, and as far as he could tell it was capable of being restored despite its very forlorn appearance, with no chimney, roofless canopy, supports bent and tyres in shreds. With that assurance, I haggled over the price to reduce it to a level that would not be ruinous if I found on getting the hulk home that it was too unsound for any future but the scrapyard. But I tried not to think what the family would say when they saw it arrive at Bressingham, and tried also to keep my gaze from a Ransome's crane engine that was also for sale, along with the rare Davey Paxman make Traction.

Scrapyards, I was finding, were depressing places to visit. So much junk covered the ground, and if here and there something of possible interest could be spotted one was hampered from even asking about it, knowing that it would involve so much heaving to get at it that one felt under an obligation to buy it. One had to try to be as hard-headed as the owners of such yards; not like my type, which easily betrays to an experienced dealer whether or not an article is really wanted. A shrewd dealer sums us up, and is liable to fix his price accordingly.

But here time was running out and I was a hundred miles from home. In buying Black Prince I'd accepted a challenge all right. It was so rough that only complete dismantling would offer any hope of complete success, and even then luck as well as toil would be needed.

Comments from the family proved to be quite as caustic as I expected. There was no doubt now, said Bridget, that I was mad; and now some Sunday visitors were heard to say it was only fit for scrap. But Jack Clements's approach was different. It was on his skill that so much would depend, and if he could tackle the mechanical side then enthusiasm and elbow grease would do the rest, no matter how long it took.

At the time Black Prince arrived, Jack and I were finishing off the Marshall. For some weeks it had stood in the barn. The wheels were removed, along with the boiler lagging, and as Ben, Don, Percy and I at odd times scraped off the grease and dirt, Jack reassembled the motion. With boiler newly lagged and cased in, painting began and ended in a complete transformation, as far as appearance went since the Marshall livery of crimson-maroon was faithfully carried out. Son Robert tried his hand at lining out, and made a good enough job to be asked to do others as well. Roger, the tubing expert who sometimes came to help in the evenings, suggested a complete new set of boiler tubes. I settled for just a few, thinking to save both time and money, but events proved that had I taken his advice, time and money would have been saved in the long run, and a bitter disappointment would have been averted.

This was the year in which the first of the Raynham Days was held at Raynham Hall near Fakenham. It was an exhibition rather than the usual rally, and included a fair and a vast range of interesting exhibits, as well as over forty engines. We decided to take both the Foden and the Marshall, and though the latter had to be transported, Boadicea ran the forty-five miles under her own steam. With Roger as driver she made good time and gave no trouble. But not so the Marshall. She was waiting for me in Dick Joice's yard at Hall Farm when I arrived there early on the first day of the Event, and feeling elated at the prospects of an enjoyable day I lit her fire and polished her up. Several other owners were likewise engaged on their engines, since this yard was the assembly point for all engines brought by low loader.

Pride and enthusiasm had reached a pretty high level with me as the Marshall responded both to polishing and stoking. After oiling round I opened the regulator when 50 lb. showed on the dial, and let her slowly tick over. The thumping noise in the crankshaft bearings had vanished, thanks to Jack. Even though he'd said he wasn't too sure if it had been permanently cured, all was set fair, I thought, for a most enjoyable two days. Then disaster struck. First came the tell-tale white smoke from the chimney, followed by an increasing drip of sooty water from the base of the smoke box door. Then with pressure at about 100 lb., out came steam as well with a sickening rush as the leak of a faulty tube widened. I could have wept with disappointment as I shovelled out the fire I'd so lovingly built up in readiness for the journey to the Park a mile away.

'Hard luck!' said the other drivers who came to stand and stare. One said, 'Never mind—I'll tow you there.' I could do no other than accept with thanks, but the ignominy of being towed was so hard to bear that I might have seemed ungrateful. I supposed it was better to be there with the others, even if my engine was dead, than leave it there in the yard out of shame. Of course Roger was justified in reminding me that he'd wished I'd let him replace all forty tubes instead of half a dozen, but this was not, I felt, either the time or the place for having such a lesson rubbed home.

For most people, the first Raynham Days were a huge success, although several thousands who had set out to enjoy the Sunday outing failed to get in. Unexpected bottle-necks had grown and queues built up for four or five miles long till many gave up. It was by far the most impressive array of engines and organs ever seen in East Anglia, and its success was not only a tribute to Dick Joice, who organized it, but a clear indication that people from all walks of life are irresistibly drawn to exhibitions with steam and other nostalgic flavourings.

Gate receipts at Bressingham that summer of 1962 had more than doubled and, after we had paid for advertising, £1,629 was sent to a variety of charitable causes. Those organizations whose local members gave voluntary help on the gate and parking, of course, received the larger shares, when it came to making the distribution at the end of the summer. But these regular open Sundays—twenty in a row—were not to everyone's liking. For me there was the joy of having an engine or two in steam, with an excuse to drive them by giving rides to children. But it was not an unalloyed pleasure for us to have people free to roam the garden. There were temptations for those who take a delight in nicking bits off plants as cuttings, and it needed someone always on the watch. Percy Piper volunteered for this duty, and though it was impossible to keep the whole garden under his eye, he did his best with the help of an occasional prowl from me. Though visitors were mostly tidy and well behaved, some would wander into the vegetable garden or the little nursery adjoining where special plants were on trial, and even into the greenhouses and conservatory. Unless doors were locked some would even walk into the house and, to Flora's annoyance, make any rest, entertaining or playtime with Anthea impossible by peering through the windows. It had to become as much a habit to secure the front of the house on a fine summer's afternoon as on a cold winter's night.

These were some of the inconveniences, much harder to put up with for the rest of the family than for me. Bridget's weekend visits became fewer during the open season, because the peace and privacy of the garden as a change from London life had gone. I can see better now how much cause there was for the family's complaining, but at the time I was much too preoccupied to fully understand their point of view. My argument was that if we were open to the public on twenty Sundays in the year, there were thirty-two we could have to ourselves and friends of our choice, and it was worth such inconvenience to be able to do some good with the proceeds. The rather plaintive suggestions from Flora that some improvements inside the house were needed were not so easy to gloss over. She had spent most of her own money already in that direction, and yet I could not bring myself to divert my personal income from writing to anything else but engines. I had written another gardening book and dozens of articles the preceding winter, but Flora upheld my right to spend such money as I wished.

Jenny arrived that October, and with two babes in the house we felt content. At least I thought I would be content, with Black Prince as the major restoration task for the winter evenings. I had no intention of buying any more engines, until out of the blue in December came the offer of another which I could not refuse. It was not only the ridiculously low price, but the fact that it was a single-cylindered Foster. With Bertha a single-crank compound and the Marshall a double, it really needed a single cylinder to complete the range in traction engines, especially as it was of a different make. Jack came with me to look at it, and pronounced it sound as far as he could tell. That was enough, and though I was somewhat torn it was so obviously a bargain that I had to have it.

14
Black Prince Rides Again

Black Prince had stood in the yard as an eyesore for weeks. It was all very well, to begin with, to hear people's comments on its shockingly derelict appearance, and to give assurances that in due course it would be restored. I would have preferred it to be tucked away out of sight after the first week, but there was nowhere to put it, because all available barn space was needed for grain storage until late autumn. With more land recently acquired, storage space was in any case inadequate and more had to be provided, but it made sense to me to add on to the existing barn where Bessie and the Foden had been restored the previous winter. The high walls—of old ship's timbers mounted on brick—could take a lean-to roof down to the stables, and take in a small partially covered cattle yard which in any case was in danger of collapse. With some scheming, in which Jack Clements again proved his adaptability as well as agility, we made a most useful addition at a very modest cost.

But for some weather-boarding at one end, all was of second-hand material. Telegraph poles and iron lamp standards came in for beams and stanchions, asbestos sheets for the roof, and a couple of large windows came from a dismantled school. By harvest time the place was ready, but once the sacks of grain had gone a few weeks later, Black Prince was nudged in backwards. There it could stay out of sight, but for us much more favourably and comfortably placed for work. With sheerlegs rigged, all the top works were lowered down to the floor, including the solid fly-wheel weighing about six hundredweight. With jacks and blocks the wheels were also removed till nothing was left but the boiler, firebox and axles. There was no doubt this time about the tubes, and with these knocked out as well, we could at last make the vital tests for boiler soundness.

Such tests are a basic necessity with something so long discarded, and what might seem primitive methods have to be used. Jack knew from long experience which areas of the firebox or tubeplate might have become thin, and could judge the thickness with the tap of a hammer. Two hammers had to be used for testing the stay bolts which hold the shell or water area of the firebox apart. There were dozens of these, and with one of us inside the box pressing a hammer firmly to each stayhead in turn, the other gave it a smart tap from outside. By this means the existence or otherwise of the unseen bolt itself could be felt. It was on those staybolts that so much depended when in steam, and as nearly all those on Black Prince were intact, hopes began to rise. Some welding would be needed where metal wastage had occurred here and there, but at least the expense of patching was not one I need worry over. There was enough to do without that, as we set to work in the evenings, scraping off rust and trying to get down to the bare metal on this old hulk of an engine.

A well-wisher had presented us with a lathe, saying this was a token of appreciation for the work we were doing on preservation. It was old but serviceable, and Jack was as pleased as Punch with it, having a machine tool that would trim and bore metal the way he wanted it. Piece by piece the jumble of working parts were brought back to polished steel with bearings and the like made good, under Jack's old but capable hands. These were tasks no one else on the place could have done, and though he put in only twenty-four hours a week, the output from his skill kept pace with the simpler but grubbier tasks which Ben, Percy and I undertook on most evenings.

The New Year was not very old before severe frosts set in with biting winds that found every draught hole in what was now the workshop. Percy slit open paper meal sacks to cover the temporary wall on the cold easterly side. Outside the ground became iron hard and the water supply to the farm failed because of frozen pipes, even though they were eighteen inches deep. But for the lack of snow it was the 1947 winter all over again, and it became increasingly difficult to find jobs for the regular nursery staff. For the first time ever a few younger ones had to be stood off, as February brought even colder weather. One or two were drafted into the workshop as an alternative to laying off, and were set to work on cleaning down the big Burrell Roller, ready for painting. This had scarcely been touched since it came, and because it was decidedly useful on nursery roadways it was well worth taking care of.

When at last, in the second week of March, the frost gave way, Black Prince was nearer completion than would otherwise have been the case. The long frost had allowed me, especially, to spend time that would otherwise have been taken up in the office and garden. Counting two or three applications of filler where the metal was pitted, the tender, belly tank, wheel spokes and flywheel were given eleven coats of paint. And the final coats were black, in a flat surface so that Bob Rolfe could make the best possible job of lining and lettering. It came as one final stroke of luck that such a craftsman lived near by and was willing to do the job more for love than money.

My luck and enthusiasm, which others had shared, had been linked with the skill I did not possess, and had produced an impressive transformation. When the great day came to put Black Prince in steam once more after twenty-two years, the hours of toil in the cold of that grim winter were forgotten, as he moved majestically out into the open. It was a proud and exciting moment for us all. The copper chimney which Flora had burnished was so high that we had to dig shallow trenches in the gravel floor to enable it to clear the sloping roof. With Whitsun not far off, there was no doubt as to which of the engines would attend the 1963 Woodton Rally. It could only be Black Prince, and since Roger Garnham had offered to drive, I was happy enough to be steersman. As there were both springs and rubber-tyred wheels it was, by comparison with last year's journey on Bessie, a comfortable ride. With 160 lb. pressure, we could keep up a steady eight to ten miles an hour, though at that speed it did not do to study the wild flowers by the roadside. Eyes had to be glued to the near-side front wheel, keeping it a couple of feet as nearly as I could from the verge. The rear wheels, six and a half feet high, were broader, and the other road users had to be considered by the man in charge of this eighteen-ton monster with chain steering.

We reached Woodton in good time, for with belly tanks we could travel eight to ten miles without stopping for water. A prize was awarded to us and success was sweet. But still I was not satisfied to let it go at that, and be content, for I'd heard of another type of engine for sale. It was a Sentinel, quite different from anything else at Bressingham with its vertical boiler, and for several weeks the thought of it niggled till at last I had to go and see it. 'Only twelve like this were made,' the owner told me. 'Even in 1930 they cost £2,500, and there's only one other left in the country.' He went on to explain that as a timber tractor it had two engines working from the one boiler. One was for propulsion capable of a speed of 35 m.p.h. and the other drove an immensely powerful winch that would haul and load large tree trunks. He had bought it in a dismantled condition, and as he lacked the time to put it together again he had decided to sell at what I thought a very reasonable price, especially as he was willing to let a slice of it remain unpaid till we'd proved it was as sound as he believed it to be.

This gentleman was among the first in the country to go in for steam purely as a hobby. One reason why he wished to sell the Sentinel was that he had come across a traction engine of 1880 and had set himself the task of restoring this to be exactly as it came from the makers, and mechanically perfect. Though our restoration standard at Bressingham had risen, we could not aspire to his level. In the course of an interesting talk he recounted how in 1940 a showman had begged him to buy a complete set of Gallopers with engine and organ because he had to go for military service. He could have bought the lot for £200, or the engine alone for £20, but lack of space and confidence made him turn it down. 'That engine now', he told me, 'would have been worth £2,000, because it was one of the finest on the road.'

It took time to reassemble the Sentinel. The boiler and tubes were in good condition, but time and again Jack and I tried in vain to bring the two halves together with their dozens of studs which had to fit exactly into their appropriate holes. Up and down with block and tackle on sheerlegs it had to go before finally it was right. After that it was plain sailing, and a trial run proved all was well. Robert had taken rather a fancy to this strange-looking vehicle, which we named Brutus. He painted it blue and white, and under Roger Garnham's guidance took it out for the first time. It had lorry-type steering and the driving position was right beside the vertical tub of a boiler. Though it weighed eleven and a half tons, it rode on pneumatic tyres and in motion made no more noise than a gentle 'push-push' sound. With Robert and Roger Garnham so keen to drive, and with it being so much more complicated than others we had, I made no attempt to master it.

That autumn Roger came on the staff. Though he lived in Diss his employer's base was over twenty miles away at Dereham. Contracts for road rolling often took Roger much farther from home, and quite often he would be away for a week at a time, using a living van as lodging. Roger, like Jack, was adaptable, and with five tractors and other farm machinery to maintain as well I was glad enough to hear him say he'd like a job nearer home.

By the end of 1963 it had become a rather stale joke to say, when another engine was offered, that this would make the collection complete. The family had by then given me up as a bad job, knowing that however much I meant it to begin with, it held no meaning now. It was an equally lame excuse to speak of my acquisitions as investments, because the chances of my wishing to sell one was obviously so remote. But I surprised them once that year. The convertible Roller—one of the three bought from Mr. Palmer—was still rather an oddity, with its spindly front wheels matching neither the heavy rear rolling wheels nor the engine itself. Then one day a youngish man called to ask if I knew where he could buy an engine. He said he lived near Colchester and worked on the railway as a signalman, and for years had yearned for an engine to play with. Within ten minutes the ten-ton convertible was sold, and the buyer was not the least bit bothered about the front wheels so long as the rest was in good order—which it was. In one way I was loath to part. This engine already had family associations. Bridget had ridden with me on it several miles along country lanes, and it was the first one that Flora had driven solo, but it made sense to let it go.

I made a modest profit on the transaction and the signalman had expected, he afterwards confessed, to pay well above the figure I'd asked. But that sum went a long way towards the price I'd been asked for a Garrett which I'd seen tucked away in Suffolk not long before. The owner had set his price, saying he was in no hurry to sell and it could wait a few months if I wished. I'd been waiting and saving up, and now it could come as yet another that would help in making the collection complete.

Garretts of Leiston had been an unlucky firm, but their 4 N.H.P. general-purpose tractor was one of their few engines that met with success. They were nippy little machines with compound cylinders, and though only weighing six tons could haul heavy loads or drive a threshing set. It was said that a German had brought out their design as a blue print, in less than a week's work.

Now those remaining were in demand because of their lightness and speed—of six or seven miles an hour, just right for those who enjoyed taking an engine to rallies.

This one, which we named Bunty, had never strayed far from the little Suffolk town where it was built in 1924. It was in good, clean condition and even had the original makers' transfer crest on its chocolate-coloured boiler barrel, as when exhibited on Garretts' stand at the 1924 Smithfield Show. For several open Sundays of 1963 it was in steam, but since its little tender was too small for passengers it gave rides by towing a trailer. It had arrived in time for the first of the rallies held at Bressingham as a second event for the Norfolk Steam Engine Club. It was a two-day affair on a Saturday and Sunday at the end of July. With ring space restricted, competitions were planned accordingly. There was no room for racing, and in any case I took the view that traction engines were not meant for speed, and could be dangerous where there were crowds.

The Saturday brought no throngs, and the gate receipts were disappointing. The Sunday was much better, and though a thunderstorm came in mid-afternoon to spoil the proceedings, the nett proceeds were a decided boost to the charities fund. A dozen engines, apart from those already at Bressingham, made quite an impressive display on the three-acre meadow as they paraded round in steam, headed by Black Prince to which the judges had awarded the prize of a Challenge Trophy presented by a club member. The parade was followed by competitions in driving skill, open to all drivers on the Saturday and to veterans over sixty-five on the Sunday. It was a sedate and modest affair as rallies go, but as a first attempt was sufficiently encouraging for us to plan for another next year.

My book *The Bressingham Story* had been published earlier in 1963, and royalties from this as well as other books and articles were proving more than merely helpful. Extravagant living had no appeal for either Flora or me, but by now I no longer cared if anyone thought me extravagant over steam engines. They were bought out of earnings apart from the business, mostly by writing at night, and I was quite convinced it would be money well spent, so long as I tried to keep a sense of proportion. With a hundred or two more in the bank at the end of 1963 than I had at the beginning, I kept a fairly sharp eye open, and almost sprang to the telephone when I saw two engines advertised one Saturday morning in the *Eastern Daily Press*.

Within hours we were looking them over. One was a Tidman Centre engine, a quite small affair used to power the old-time roundabouts, and it even carried a little donkey engine on top, which drove the bellows for the organ. Someone had converted a set of roundabouts from steam to diesel, and here was a Norwich-made engine as pretty to watch in motion as one could wish to see, with plenty of brass and moving parts. Having no wheels on which to run, it had been mounted on a low trailer by someone who just wished to take it where it could be seen for the charming little machine it was. With the certainty that it would be admired by visitors to Bressingham, and having been told by the owner that others had telephoned in response to his advertisement—this I later found to be true—a bargain was very quickly struck. And a bargain it proved to be when it came to later developments in which the engine played an important part.

15
The Little Iron Road

The other item offered with the Tidman Centre engine was a model locomotive. It was for $7\frac{1}{4}$-inch gauge track, of free-lance build, not to scale and of unattractive appearance. But if this in itself aroused no fresh hankerings, it served to resurrect some I'd been stifling for a year or more.

I had developed the habit of putting out feelers to the family when acquisition urges came. Robert, Adrian and Flora could not be expected to offer more than a certain tolerant blessing on my enthusiasm and this, so far, they had given. But if I carried it to what they considered to be an excessive degree, harmful to the future well-being of the place in which they too were deeply concerned, it would have been a mistake ever to override their opinions. The future of the business side, especially, lay with them rather than with me. But prospects for the future of the business would remain bright only if family ties remained firm.

My idea of installing a railway had been mooted almost a year ago. The four of us were returning from one of our rare trips to Norwich for a meal and the cinema, and were feeling the better for it. It seemed a good time to sound them. I'd been toying with the notion of a small locomotive and a track for it to run on, though I was not sure where it could run. The idea had emerged from three lines of thought. One of these was connected with a growing fear that giving rides to children on traction engines ought to stop. An accident could occur in which a child might be hurt, and in any case the increase in visitors was restricting running space. The second thought was more mercenary, because with a little passenger-carrying railway, fares could be charged which need not go with the gate money for charity. It might even show a worthwhile profit in time. And the third factor, apart from the interest in steam, was the news that a large North Wales slate quarry was then selling off its narrow-gauge locomotives at very low prices to steam enthusiasts.

In a roundabout way I made all these points, but neither wife nor sons were impressed. The objections they raised were valid enough to them—that the peaceful atmosphere of the garden which meant so much would be spoiled; that wherever sited on the nursery or farm, the railway would have a nuisance potential; and that I had enough on my plate as it was. Perhaps their minds went ahead of my proposition, which to me involved only one small locomotive and a few hundred yards of track. I was disappointed but tried not to show it, for I had to give them credit for being reasonable and tolerant people, and if this was their reaction then I'd best forget the idea.

For a time I did manage to push it out of my mind, but after a few months it began to nag again, though I kept it to myself. Flora had an inkling of the way I felt when we saw that $7\frac{1}{4}$-inch model for sale with the Tidman. She was not taken with it as a working model any more than I was, but at least she did not condemn the hints I let drop that a railway of about this size, with a more attractive locomotive, would not impinge on the garden if it were placed well to one side.

Spring of 1964 was awakening, and for the next few weeks as usual the garden and nursery claimed every daylight hour. And it was work I still enjoyed, despite my overgrown hobby of steam. Even visitors such as Mr. Stebbing, who occasionally called in out of sheer interest in steam, had to put up with my switch of interest if it was spring. But one day in March Mr. Stebbing came to tell me that if I wished to have Cissie to partner Bessie again, he would be glad to let her come at much less than he'd paid for her. The trouble, he said, was that working alone on her restoration had proved too big a job, and for nearly two years she had scarcely been touched. She was partly dismantled but could be towed in, if I would have her, and if I cared to pay a few pounds as deposit, the remainder of the £100 purchase price could wait a year if need be.

Having paid the £10 deposit I mentioned my interest in a miniature railway, just in case he knew someone who might have one for sale. 'It's funny you should ask me that,' he smiled, 'because only a week or two ago an old friend of mine said he wanted to sell one he'd spent years in making. Miniature locos are not much in my line, but I can find out more about it if you like.'

I did like, and a few weeks later, when the spring rush was easing, we went one evening to see Commander Robinson, near

Bury St. Edmunds. He apologized for not getting out of his wheel-chair, saying that for the past ten years he had had no choice but chair or bed. 'And for ten years', he explained, 'I've not been able to touch my little Tilbury Tank. It took me twelve years to get within a few weeks of finishing and then I had to leave it. I'm seventy-nine now and realize I never shall finish it. I want it to go to someone who will clean it and paint it in the correct apple green livery, and if it can be used on a track so much the better. You go and look at it, and then come back and tell me what you think. My housekeeper will take you to the shed which used to be my workshop.'

Inside the shed, on a specially made stand, covered with oily sacks and canvas, there it was just as the Commander had left it. It was a 4-4-2 tank, about five feet long, and of about 7-inch track gauge, almost complete, and as far as I could tell, a perfect scale model of the famous Whitelegg locos used on the London, Tilbury & Southend Railway. It was rusty as it had never been painted, but less so than one would expect after being left for so long. Back in the house the Commander named his price. I had already gained some idea of current values, and this was so obviously a bargain, I wrote out a cheque there and then. Even if it proved too small to pull little trucks in which children could ride, it would be worthy of being on show to visitors as a model. And I would enjoy the task of rubbing down and painting such a charming little engine.

We were about to leave when suddenly the old Commander said he wished me to have something else he'd left unfinished. 'It's a box full of parts for a model marine engine based on an 1881 Cunarder. You have them, because my time may soon be up, and I'll be happier to know that at least they won't get sold for junk.' Merely to thank such a man seemed inadequate. I had no idea of the value of these marine engine parts, but that did not matter. What impressed both Flora and me was his realistic philosophy of life. He had hung on and learned to live with a crippling, incurable disease without bitterness or complaint. I found myself thinking about him, and how I might face such a situation as his, for much of the time that I spent on putting the finishing touches to the little locomotive he had so cleverly built. But by the time this was finished, I had gone a step further towards installing a working miniature railway. Someone with experience of miniatures had seen the 'Tilbury' and explained that as a model it was perfect, but would not be very practical for what I had in mind. The driving wheels of an exact scale model of a passenger locomotive mostly proved too large, making for excessive speed but diminished starting power. I could appreciate this and could foresee difficulties and disappointments if, as would be the case here, a lot of children wanted rides in the short space of time of one Sunday afternoon a week.

This counsel was disconcerting, even if it made sense. It would obviously be silly to buy 7¼-inch track to prove the point, and in any case, track sufficient for a worthwhile run would cost more than I'd paid for the engine. Since its arrival I'd been taking the *Model Engineer* and now began scanning its 'For Sale' columns, because the more I thought about it, the more determined I became that a garden railway must come, and the sooner the better. Late in May, following a day at Chelsea Flower Show, we went on a little round of nurseries and gardens, and on a Sunday morning called at an address at Bexleyheath. There we were met by Dick Simmonds who introduced us to Mr. Hamment, the owner of some locomotives for sale. He had been in the building trade but had amused himself in recent years by running a miniature railway in nearby Danson Park. After falling out with the local Council he had taken up the track, and was off to Bermuda as soon as his affairs could be cleared up. As a local professional model engineer, Dick Simmonds had been roped in to find buyers for his 9½-inch gauge railway equipment.

The sight of a red 'Princess' Pacific set up an immediate yearning, which put as much strain on my reason as I knew it would place on my resources. If ever I wished that cost need not deter me, this was the time, for there was a Yankee-type loco as well as five trucks and several hundred yards of Jubilee track stored elsewhere, all for sale. Mr. Hamment said he would prefer to sell the lot together. 'I'm willing to split,' he added, with what I took to be a hard glint in his eye. 'But you can't expect to get so cheaply whatever it is you want if I have to do that.' It was not a good atmosphere in which to do a deal. Correspondence had been with Dick Simmonds, and now it seemed Mr. Hamment was taking over, despite this agency. I sensed that relations between the two had become strained, and when the figure of £600 for the 'Princess' Pacific was asked, I saw Dick turn away as if he wanted no further part in it. This prompted me to haggle, and in the end Mr. Hamment gave way, and for a lower price than £600 he included a hundred yards of track, and some bogie trucks. A week later it was at Bressingham. 'Whatever have you got there?' Jack Clements said, beaming all over his lean face. 'That's a little beauty, that is!'

So it was, even though it was not built to scale, since its wheels were disproportionately small. It had in fact been built for just the purpose I had in mind and, as a 4-6-2 Pacific, it was near enough to the L.M.S. 'Princess' design to satisfy me. I was indeed more than satisfied on other counts. There was room to sit in the tender and drive with feet on the footplate, and even if the gauge was only 9½ inches, the length of engine and tender was twelve feet, and the weight close on a ton. It was June, and for a week or two I dropped everything else to lay the track. Some of this needed repair, and odd pieces of wood were rustled together for sleepers. Taking a course between the garden and a shelter belt of trees bordering the road, the line began on one side of a parking meadow. I made it curve in under some overhanging hollies, which provided a tunnel effect and broke up the straightness, and until we made our first trial run I believed it to be level or thereabouts.

It was prudent, I thought, to let Roger make the first trip. He had taken such a fancy to it, and in any case I wanted just to lie on the grass and watch its passing, with its fascinating Walschaerts valve motion in action. A juvenile urge maybe, but I thought this miniature would be the nearest I would ever get to the real thing. The 'Princess' chuffed merrily past a few times and, though Roger reported that the steam brake would not work, he said he could hold the engine on the reversing lever. It wouldn't matter, we thought, to use this for stopping until someone with the necessary skill could repair the faulty steam brake. Then my turn came to drive, but in spite of the enjoyment this gave me the track was far too short.

The following Sunday there were children wanting rides, but to begin with I hadn't the neck to charge for taking them the hundred yards to the big block of wood that served for buffer stop, and shunting back to the starting point. The first load I took nearly ended in a minor disaster. It was not until we ran with a load that I realized how much gradient meant to a locomotive. The 'Princess' barked somewhat at the beginning of the run, but once through the holly bush tunnel it gained speed even with the regulator closed. With more enthusiasm than knowledge I thought it safe to shut off steam, and then pull back the reversing lever for the last twenty yards so as to stop just short of the buffer. But this, I found to my horror over the last few yards, was not checking speed enough. In that moment I could think of nothing else than to sit tight and hope for the best, while holding back the reversing lever. It simply didn't occur to me to open the regulator and put the motion in reverse under steam. Over went the big block of wood, which I'd omitted to tie down. As it happened this was fortunate, because the leading bogie cleared the end of the rails and so did the first pair of drivers, and came to rest on the block. Only a little jolt to the passengers and no scare, but it scared me as I reflected what damage there might have been had the buffer block been firmly held down.

It was not long before I came to terms with Mr. Hamment for some more tracking. Luckily for me he had sold his other locomotive to someone who did not want the rest. To lay this track on a properly graded bed would be a job for winter days, and indications were that this little railway would prove to be a good investment. One pointer to this was the interest shown by many well past childhood. Some would just stand and stare as the little locomotive ran back and forth, but others would squat with cameras, hoping especially to take pictures of it emerging into the sunlight. A few even brought their tape recorders. At this time the beat revealed faulty valve settings, but Dick Simmonds offered to put this right before another season began. He, by the way, had decided to retire and having asked me to help in finding a house with an acre of land for his own model railway, had moved by the end of the year to Wortwell only twelve miles away.

The 1964 Rally was a real success, and swelled the Charities Fund by £1,232. It was held on August Bank Holiday Sunday and Monday, and on each day some 8,000 people paid at the gate to see the show of engines. The competitions were more varied, and a tug of war was introduced. This was fun. The first engine used on the rope was a Garrett Tractor, and as more and more people ran into the ring to try and hold it, gradually they gained the mastery. Then out came a larger engine—Leslie Aldrich's Burrell Compound, which he'd bought from Tim Stevens. Leslie pretended for some minutes that the long line of about seventy or eighty people—men, women and children—had him held. Then with full throttle and barking exhaust he slowly gained ground. As the wheels began to churn up the grass, it was time for me to declare it a draw, rather than let the engine dig itself in.

16
In with Both Feet

It had surprised me to discover how few members of the Norfolk Steam Engine Club seemed to be interested in railway locomotives. The conversation at its spring social evening and autumn dinner following the A.G.M. was interesting enough but entirely confined to road-using engines. Soon after I became a Committee member in 1962, I learned that the one remaining 'Claud Hamilton', perhaps the most famous of all Great Eastern locos, was for sale at £1,500. This was the scrap price and Mr. George Goodings, the B.R. Divisional Manager, had said if it was not soon sold for preservation then it would go for scrap. The Club had good reserve funds, but when the suggestion to buy came up, only one other Committee member was the least bit in favour, and the majority were against anything but keeping its reserves lest a rainy day led to a loss on the Woodton Rally. Traction engines were the thing. They had vanished from the everyday scene but it was still not realized that nostalgic yearnings for railway locomotives must also follow their disappearance.

This was one reason why I began to pine for a full-scale locomotive. The other reason was that, far from being content with the miniature 'Princess', I found that this too made me wish heartily that I could afford a full-sized one. With hopes rather dim I asked Mr. Goodings later in 1964 to find out what British Railways might have, retired from service, for sale at the nearest depots to Norfolk. This revealed that steam was very quickly being pushed back, and even March and Peterborough had gone over to diesel. I was booked to give three talks on successive nights during that November, so I went by train to Hull for the first and called in by appointment at the Doncaster loco depot. There they stood, dirty, discarded and utterly forlorn—engines that had been so proudly and soundly built, and long before their time were to be slaughtered.

I looked inside the smoke box of a Thompson B.1, and climbed on to a V.2, mixed traffic engine, and longed rather hopelessly to save one and to give it an honoured resting place at Bressingham. These heavier engines were then valued at £2,500 or so, and that kind of figure was out of the question for me. At Colwick three days later, only three engines were for disposal, and these too were well out of reach in price.

It had been a disappointing trip, not only because of the sight of those locomotives I had to leave to their fate, but because I'd gone by train. The journey to Hull via Doncaster had not been difficult, and I could not blame the shorn B.R. services for the mistake I made which landed me in Hull a day too soon for my lecture. Having changed into evening clothes I went to the Hall where I was to give the talk, only to find it locked and empty. This put me into a panic until I checked to find that I'd not, as I feared, got my three engagements in the wrong order. On leaving Hull, I'd imagined it easy enough to take a train via Doncaster to Mansfield, where my next talk was to be. But there were no trains to Mansfield, nor was there a service from Mansfield to Nottingham, and though I prefer train to car travel, and car to bus, there was nothing available but slow bumpy buses. After I had given my talk at Nottingham in the afternoon I found that no train service could get me beyond Peterborough that night, and I had to telephone Robert to come and collect me the hundred and ten miles by car.

At home again I studied once more the price list of locomotives Mr. Goodings had procured for me. The cheapest was £1,200, and that because it was the lightest in weight, and I would have to borrow the money on my firm's account if I decided to buy. For days I dithered, reluctant to drop the idea despite its patent impracticability. I paused often on my way to the office to conjure up a picture of a loco standing there as a background to the traction engine yard, on rails just long enough to support its wheels. And then slowly came the realization that if one were there, if all the obstacles of cost and transport could be overcome, it would have to remain a lifeless thing. No matter how much interest it would give to visitors, no matter how glad I felt at having saved it from destruction, it could never be satisfying without fire and water to bring it to life, and a track on which it could move.

Traction engines have this advantage over railway locomotives. They can be brought to life with fire and water even if they do not travel at all. I had recently seen one which had not moved a yard for about thirty years. A man from near Halesworth had telephoned to say that he had an old Marshall Portable, and it had to be moved quickly, either for preservation or for scrap, and did I wish to bid for it? So off we went, on another of our jaunts. Flora, Anthea, Jenny and I walked with the owner up a very narrow lane to the remains of a windmill. Nearby was a tumbledown shed and inside stood the engine, gaunt, rusty and with the smoke box caved in. The owner explained in a breezy sort of way that until the war, the Marshall had served to keep the mill going when there was no wind. After the war, there was no need either for the mill or the engine, but there both had stood decaying ever since.

As it was a portable, and in such a deplorable condition, I felt little of the usual urge, and almost wished I'd not bothered to come and see it. But having seen it, and knowing that if I did not buy a scrap merchant would, I bid £20 which I reckoned to be scrap price. I'd taken a quick look inside the firebox and by striking matches had been able to see that some tubes were missing. The rest were pretty rusty, but it could at least be stowed away against the time when, with other restoration jobs done, we wanted something else to tackle. Jack Clements had in fact expressed some fears to Robert that he hoped I'd see to it that he wouldn't be stuck for a job in years to come.

The owner tried hard to wheedle a higher offer out of me, but for once I was able to hold out. In due course the engine arrived, and as usual Jack made a close examination in the open. When he'd finished he gave a slow smile. 'One thing about it,' he said laconically. 'That'll keep me going quite a time, if I should run out of a job. But I reckon there's nothing missing except a piece inside the box, just above the firebox door. Somebody cut that out to put a patch in, but never did—I expect you saw that?' I had to admit I'd not seen it. Only a torch held from inside the firebox would have revealed it, in the darkness of its shed. Back to darkness in a recessed part of the barn it went, and there it has been ever since, because Jack has had plenty to do without tackling that.

The next purchase was not so much of a pig in a poke. A Norwich firm of nurserymen had an engine they used for soil sterilizing and it had just been rejected by the boiler inspector. The tip was given by Walter Page, who was to follow me as President of the Norfolk Club, and who had presented the trophy to be won at the Bressingham Rally in the form of a strangely beautiful brass oiling contraption. Mr. Page said the engine was a Fowler single cylinder and reckoned it was at least seventy-five years old. Jack came with me to see it, and as he poked around expressed some surprise that the boiler inspector had not rejected it long before. As he used his hammer to test the thickness of the tube plate, it went through in one place. 'Here's the cause of the trouble!' he exclaimed. 'Mud—just mud accumulated inches deep above the foundation ring. Ruin any boiler, mud will, if it isn't swilled out regular. Turn dirty water into steam and you get mud left behind, where it can only do harm.'

But because it was so old, because Jack reckoned the cylinders and motion gears were good—since use had been largely confined to raising steam for sterilizing soil—I had to buy it. The price was low enough anyway, and I was glad the owner had remembered a small favour I'd done him years before when supplying a collection of plants for his own garden. There was no doubt as to which of these two old crocks should come in for restoration as a task for winter evenings. Ben, the pigman, had become (with me) the remnant of those prepared to strip down, clean and repaint an engine just for love. He had developed quite a technique, but varied it according to need. If caked oily dirt had to come off, he used a putty knife, and then brushed down with diesel oil till clean metal emerged. But sheer rust called for an electric emery disc, and the air became so thick with dust that a mask and goggles would have made it less of a nasty job. But Ben seemed not to worry about such things and learned to keep his head clear of the cold dusty blast. His was the donkey work, and as I followed him up with scrape-on filler paste for the pit holes in the metal, we made slow but steady progress as a pair.

We decided to go the whole hog in 1965 as far as visitors were concerned. It was my decision more than anyone else's, but even I had been somewhat half-hearted about this aspect until then. Flora had become fed up with the lack of privacy on Sunday afternoons. I did not blame her for not wanting to show Sunday visitors round the garden. She had done more than her share of that in the past few years as well as gate duties on Sundays, and in any case the care of Anthea and Jenny ruled that out. Sunday visitors were increasing in numbers and because so many seemed disappointed that there were no plants or flowers for sale, Flora came up with the idea of running a stall. A refreshment stall had already been run by the Bressingham 'Over 60's' Club until it was too much for them to cope with, even if they found the profits useful. In 1964 provision of teas was handed over to an outside caterer, which left the 'Over 60's' out until 1965, when we decided to open on Thursday afternoons as well. This was how the pressure was building up. Not everyone who wanted to see the garden could turn out on a Sunday, and though the 'Over 60's' were willing enough to do the catering on Thursdays, our own staff had to man the gate and parking arrangements.

All this entailed the need for more advertising, and since we had noted the influx of people following some little feature of the place on a few television programmes, I decided to include a few pounds' worth of seconds once a week on Anglia TV. The extra had to be paid for, and the entrance charge had to go up from 1s. 6d. to 2s. for adults. Anticipating larger gates brought the need to arrange for more Sunday volunteers for gate and parking, and other local charitable organizations were added so as to spread the duty rota. Coach parties had to be by appointment only, except on Sundays, to avoid muddles and because of the need to keep the place free of visitors for some days during the week.

So it was in 1965 that, for the first time, some direct gain came to the business from visitors. Gate receipts were earmarked as usual for charitable causes, but now those from the little railway, as well as the stall, went into the business account. It was soon apparent that the 'Princess' would in time show a profit on the outlay. The track had been made into a 750-yard round trip with a loop and spring-loaded points at each end. The hurriedly laid section on which running had begun had to be almost completely relaid when a survey revealed a five-foot fall from one end of the garden to the other. And I'd thought it near enough to level. Now the 'Princess' could haul its little trucks, holding up to thirty children, but adults too wanted to ride. With Roger as chief driver, who preferred to stick to an engine he knew, I took over only as relief, but though giving rides on traction engines had ceased at least we still had two or three in steam, taking them in turns.

The one engine to be steamed every Sunday was the Tidman Centre, on its little trailer. This was not merely because it was so attractive, with so much brasswork and moving parts, but because Bert Swann had turned up as a voluntary driver. He had driven traction engines in the First War—hauling guns in France—but for years since had become so crippled with arthritis that he could only hobble. Now twice a week he drove his little invalid three-wheeler in, once to clean and polish, and on Sundays to take charge of the Tidman after Ben had lit up for him. It was as much of a joy to him as anything he could wish for, and always a little group of admirers stood around, with whom he could natter.

The new arrangements worked well enough, and at the end of the season the nett proceeds for charity had jumped to £2,764. This included those from the two-day rally, which had become a regular event. Apart from marquees for 'Floral Art' and 'Cage Birds', put on by local societies, stalls for the Festiniog and Talyllyn Railways and one or two others turned up. And by arrangement, so did Screeton Brothers with their Galloping Horses. They came a hundred and twenty miles, and though we had to guarantee them £100, the attraction of an old-time Roundabout with its splendid organ proved worth while, and people flocked to ride the prancing horses. The throng, the bustle and noise of what we termed the 'Bressingham Steam Gala' were exciting while it lasted. It was also a bit of a headache for the few of us who had to arrange it and make it work. It was tiring, too, and when Bert Swann said he hoped I'd find a set of roundabouts to go with his Centre Engine I laughed. 'Hardly likely,' I said wryly, 'I've got enough on my plate already.' This was true, but only up to a point, because by then I was beginning to cast about for additions I considered more important to me than roundabouts.

By that autumn of 1965 nothing seemed so desirable and urgent as another, larger railway and locomotives that would take visitors into the nursery which so many wished, but could not be allowed, to see. They had been barred from the nursery ever since the garden was made large and varied enough by itself. At least that was what I would have thought, but now the little garden railway had proved a profitable venture in itself, and it was pretty obvious that steam was the draw. And if a steam railway offered trips elsewhere on the place, then this too would be of interest to visitors, as well as useful to the business. The potential was there, and we might as well use it. I explained the idea once more to Flora, Robert and Adrian. This time, even though some of their misgivings remained, there was no opposition, and in any case I would be cautious when it came to outlay, and go in for a short track to see if my hunch was correct.

Part of the hunch was to do with the use to which a railway could eventually be put for the sake of the business. The nursery occupied several fields, some lying half a mile or more from the packing sheds. Every autumn when plants had to be dug up to fill orders, the dirt roadways serving these fields gradually deteriorated with constant use. As rains came the ruts grew deeper, and from November to spring were so muddy that transport became a real problem. To make up roadways with hardcore and gravel was not the answer because, apart from the cost, nursery fields were often rotated with farm crops for which weather-proof roadways would be a hindrance. It seemed to me, therefore, that if a short rail route to begin with attracted paying passengers, then there would be a cast-iron reason for going much farther afield so as to serve a dual purpose.

1. 'Bertha', the first Traction Engine of the present collection—acquired in 1961—is on the left and 'Boxer', the 8-ton Burrell Roller, on the right

2. The Foden Steam Wagon in the Museum shed

3. Fowler AA Ploughing Engine 'Cissie'. Her sister engine 'Bessie' was the first of the pair to be restored

4. The Foden Wagon in 1962 with restoration nearly completed—Alan Bloom at the wheel

5. Burrell 'Compound Portable'—believed to be the only one in existence

6. Jack Clements and the author on the Marshall after restoration, 1962

7. Jack Clements examining 'Black Prince' after arrival from the scrap yard, 1962

8. 'Black Prince' as restored to a Showman's Engine, 1963

9. Steam Rally at Bressingham, 1966

10. 'Gwynedd', the Hunslet 2 ft. gauge, on a trial run in 1966

11. The Metre gauge Peckett 0-6-0 with the author and daughters Bridget, Anthea and Jenny on the footplate, 1967

12. Burrell 12-ton Roller, acquired in 1961. Standing beside it is Sid, once a steam roller driver, who is one of the volunteer drivers on open days

13. Arthur Prentice takes 'The President' Burrell Road Loco. This is in turn with other Traction Engines on open days

14. 'Thundersley' safely on the new track

15. 'William Francis'

16. 'Oliver Cromwell' in steam at Bressingham

17. Close-up of 'Oliver Cromwell'. The driving wheels are 6 ft. 2 in. diameter

18. 'Oliver Cromwell', en route from Diss Station to Bressingham, moves with extreme care below a bridge carrying the main G.E. line, 1968

19. Preparing the track bed for the 'Woodland Railway' under difficult conditions, January 1969

20. Tracklaying in the woods, Arthur Prentice and Don Hubbard follow up with the rails on sleepers laid by Alan Bloom

21. 'Brunhilde'—the Schwarzkopf on the Nursery Line

22. 'George Sholto' approaching the loop points on the 'Woodland Railway'

23. The organ from Flora's Gallopers in winter storage

24. The runaway 15″ Krupps now righted after unloading from Germany, 1972

25. 'Rosenkavalier'—the Krupps after restoration

26. 'Royal Scot' shunting 'Duchess of Sutherland' into position, 1973

27. 'Duchess of Sutherland' after restoration

28. 'Millfield' whilst at work at Sunderland docks

29. A Sunday morning at Bressingham, with 'Bronwllyd', 'The President' and 'Royal Scot', 1973

17
Rails and Relics

The weather did not look very promising on the morning Roger and I set out for North Wales. It was late November 1965 and by the time we reached Kettering it was snowing hard. The car began skidding near Market Harborough, and with the sky seemingly full of snow I decided to go by train from Rugby. At the station they said trains to Crewe went at a hundred miles an hour, and after we had waited nearly an hour for one to pick us up, this proved to be the case. The only snag was that in putting up such a speed on the new electrified line, the train stopped here and there on the way so as to allow the time-table to catch up. North of Rugby, there was no sign of snow and only hill tops in North Wales were peppered when we finally reached Bangor. There our troubles began. We went by bus to Bethesda only to find that all the publicans and hotel keepers were out on a spree and would take in no lodgers. We found one hotel back in Bangor and next morning returned to Bethesda and there saw the remains of what had been a lively team of narrow-gauge locos of the Penrhyn Quarries. Of the five that were left, only two were anywhere near serviceable and all had some vital part missing.

But this was rather to be expected for I knew that the best had gone a year or more ago, and those remaining had been somewhat cannibalized. It was no use kicking myself for having dithered for nearly a year. At £100 each they had been real bargains, and some had been shipped to the U.S.A. for much more than this. I bought 'George Sholto', which appeared to us to be the one in best shape. I had intended buying two, but was told that 'Hugh Napier' had just been earmarked for the Penrhyn Castle Museum. This was also a Hunslet make but the rusting chassis of a Hudswell Clarke awaiting scrap intrigued me with its Walschaerts valve gear. This I bought for £30 just in case the day came when, with a boiler, it could come alive again. And to go with 'George Sholto', I bought the five ancient trucks once used to take the quarry men to their work, and three old slate wagons.

Though we had fared better than I dared to hope, so far as purchases were concerned, we were not out of the misery Roger and I had let ourselves in for by abandoning the car at Rugby. Part of our plan was to see what might be available near Cemaes Bay on the coast of Anglesey. It took hours by bus and to reach it we had to go the remaining ten miles from Amlwch by taxi. It was dark and blowing a gale when we reached that desolate spot, but by the light of matches and candles we were shown some of the fantastic collection amassed by Capt. Hewitt, a recluse who had died a few months before. Shed after shed was crammed with an almost unbelievable variety of collectors' items ranging from stuffed birds and birds' eggs to machine tools, pictures, model engines, ships and books. And nothing was for sale. The whole lot had to be put up for auction, our guide told us, once the well-nigh impossible task of making an inventory was accomplished.

Once home again there was plenty to think about and plan for. Other inquiries for narrow-gauge locomotives had led to one named 'The Doll'. This was a Barclay 0-6-0 of no great age, which had been acquired from a Midlands steelworks by a Society formed more or less for the purpose of preserving it. The Society had, it appeared, fallen apart, and the principal stakeholder thought it best to sell. A final decision was to be taken at a meeting early in 1966 and though I was promised first offer I had to wait as patiently as I could. With a television programme in 'Gardening Club' booked for early February there would, I hoped, be something else to brighten up the tedious trip to the Birmingham B.B.C. studios, because 'The Doll' was shedded not ten miles away. Word came just in time that the offer was confirmed, and at a price so reasonable that, in my mind, I'd bought her even before I set eyes on her. She was, in fact, a pretty little thing, prettier than most industrial locomotives. With Flora patiently waiting once more I spent half an hour examining what I thought to be the salient points about her, but knowing precious little how many factors needed to be taken into account even with a 2-foot gauge railway. It did not occur to me even to check the gauge, for although I knew that the slate quarries were in fact 1 foot 11 inches I had assumed that this was the 2-foot standard, and 'The Doll' would be the same. But it was not merely ignorance. The fact was that after waiting for six weeks on tenterhooks, I was once again suffering unconsciously from over-enthusiasm, with the subject needing only a 'yes' from me, and not even a request for a deposit.

Having decided on a route for the first Nursery Railway, with a much more thorough survey for levels, the next requisite was to procure the rails. Had I not delayed for a year before going ahead rails could have been bought for about £10 per ton from the Penrhyn Quarries. With the takeover by McAlpines, and a switch to lorries for transporting the slate, there were fifteen miles of track within the quarries which, with other equipment, the old company had to dispose of for scrap. All had been sold at the time of our visit, and the scrap merchant's price of £22 per ton delivered was, I thought, a bit steep. In the end, I had to accept it because other offers were even higher. In March, Roger journeyed on his own to Bethesda. The scrap merchant had said we must mark the lengths required, since some was still *in situ*, so that when delivered we could not quarrel with him. Someone else had bought a load and refused to accept delivery, and he was going to make sure we would not do likewise if some of it was bent.

Roger returned to report that a half mile of track had been daubed with yellow paint, with other colours used for points. All that remained was to wait for it to come by the scrap merchant's own lorries. Then came a shock. His secretary telephoned to say that the two loads, of about eighteen tons of rail, would be arriving in three days' time, but the price would now be £28 per ton instead of £22. I had to think quickly. We had a letter confirming the deal at the lower figure, and if I were to try at this crucial stage to force him to keep to it, he could easily wriggle out of supplying at all. 'Will you accept?' the girl asked casually, with me fuming within. 'Well it's a bit late in the day to stick the price up,' I said, 'but I don't think we need fall out about it.'

I was afraid this might bring her employer back for a more definite answer. But the loads arrived, on two decrepit lorries, the Saturday afternoon following. I indicated the spot for unloading and up went the tipping gear, and in ten minutes the lorries had gone, leaving an untidy eighteen-ton heap of rails behind.

In due course the bill arrived, at £28 per ton. By return a cheque was sent at the quoted price of £22, and then it was the merchant's turn to fume, but he had to give in and keep to his bargain.

The rails had to remain where they were till the spring rush of work was over. Some preparation had been made in midwinter. On digging a ditch, which was badly needed for drainage anyway, the soil was placed along the edge of the big pond. This two-acre stretch of water had been dug out at odd times between 1955 and 1959, but now it appeared as if it had always been part of the landscape, margined by willow trees and rushes, and with big patches of water lilies. This two-hundred-and-fifty-yard length was matched by another, one hundred and fifty yards higher up the gently sloping valley. It thus made a complete circuit enclosing the most concentrated part of the nursery, with sheds, glasshouses, frames and the plunge beds, where about six hundred thousand alpine and other plants were grown in pots. The slope was about 1 in 40, representing a drop of seven feet from where the platform must be to the pond-side stretch, and apart from easing it as best we could, we had to wait and see whether or not 'The Doll' or 'George Sholto' would climb back with a load.

With ex-British Railways sleepers sawn in half, tracklaying began early in June. Crossings and culverts had to be made, and there was gravel to lay and points to fix. It was five weeks later before the test could be made. We were amateurs at laying a railway track, but though it was hard work there was no lack of enthusiasm from the two or three who helped me. For Don and Ben in the evenings it was something different to their normal work, and for Roger it was the realization of a hope that before it was too late a narrow-gauge steam railway would be installed.

'George Sholto' was on the way to being restored. Some parts were missing, which Jack had to make on the lathe. These were jobs only he could do, and he had to take his time. In any case 'George Sholto' had been bought with a longer railway in mind, and 'The Doll' should be quite capable of coping with the 1966 traffic. The exciting moment came when with 100 lb. pressure, her regulator was opened for the first time at Bressingham. Round the track we went, trying her out on the curves and gradients, gingerly at first but with confidence growing. The eccentrics were a little sloppy and glands needed attention, but when she blew off at 160 lb. we felt reasonably happy, and hitched on the old workmen's trucks, which Ben had painted red, just as the first visitors arrived that Thursday afternoon. A coach party from my native village were given a free ride, and all went well. But on a later trip the leading truck left the rail on a curve, and having jacked it on again we found that the three-inch dogspikes had worked loose. 'The Doll' must have caused that, we thought. Reasoning it out, it seemed that being an o-6-o, with a longish wheel base, the wheels under the eight tons' weight had tended to straighten out the curves, thus widening the gauge and allowing the truck to fall in between. Believing the remedy to lie in removing the flange from the middle wheels we cut these off, and by the following Sunday all loosened dogspikes had been hammered home again, and the gauge pulled in. Guide rails, too, were placed on the sharpest curves in the hope that these would make for safety.

Sunday afternoon running began well. The trucks were devoid of springs, and the coupling chains were so slack that it was almost impossible to get off to a start from the platform without jerking the passengers. Another jolt came at the end of the first fifty yards, when applying the brakes over the sloping curve down to the pond. The passengers seemed happy enough, and took the jolts as a joke, as they did the smoke which sometimes wafted over them. But just as Roger and I, as driver and fireman, were feeling more hopeful, down went a truck on a bend, with a lurch and a list by the time we stopped. A girl shrieked, and though the passengers in the other trucks were safe enough, we were worried sick lest someone was hurt. No one was injured, but the shriek had caused a mild panic; however, some occupants of the derailed truck were laughing about the mishap, as they stepped down. Willing helpers heaved with Roger and me, and we soon had it back on the rails.

Roger stayed behind. With the circuit completed he met me with the news that the rails had spread once more, and again the dogspikes had been forced out. 'Sorry folks, no more trips until we've fixed the rails.' I said this in a cheerily hopeful tone, but without much idea how we could. Next morning came the full inquest. Still believing that 'The Doll's' rigid wheelbase was too long for the tight curves we'd had to lay, it was not until various possible remedies had been discussed that Jack suggested comparing the distance between the wheels of 'The Doll' and of the trucks. 'There's your trouble,' Jack explained dolefully. 'Full two foot between the engine wheels, one ten and a half on the trucks. Engine spreads the rails, and down go the trucks. Simple as that.'

The blame was of course on me. I had not checked it before it arrived or for that matter before deciding to purchase 'The Doll'. There was only one of two decisions to make. Either we must reduce her gauge to be compatible with the trucks and 'George Sholto'; a major job involving, as far as we could see, a rebuild, and taking far longer than to bring 'George Sholto' into service. There was only one sensible way out of the dilemma, but it was more than merely disappointing to have to take it.

Two weeks later, 'Sholto' was towed out and jacked on to the track and given a quick coat of black paint, which was just dry enough to permit raising steam next morning. In the meantime, with 'The Doll' placed on the siding, the track was trued up and fixed down with heavier, longer dogspikes. There was no doubt now that with both engine and trucks of the same gauge there would be no more derailments on that score. It was late July, the period when gates reached their peak, and there was no doubt either that given time, this new venture into steam would pay off. 'All right,' said Flora, Robert and Adrian indulgently. 'You were right and we were wrong.' I couldn't help rubbing it in a little, and having scored a vital point, I decided to go right ahead with another longer track.

Three other worrying experiences had occurred during that eventful summer of 1966. The first came just before we tried out 'The Doll'. A man from the Midlands had written a long letter full of cryptic forebodings, saying that he wished urgently to come and talk over a very serious matter. His first announcement on arrival was that I had no rightful ownership of 'The Doll'. It had, he said, been sold to me without the authority of the legal owner, and there was a cast-iron case in support of this. The true owner, whose representative he was, was prepared to help me recover the purchase price if I would refrain from dispute, and hand over the locomotive at once. If I was foolish enough to resist, then legal proceedings would ensue, leaving me the loser of 'The Doll' in any case, and perhaps the purchase price as well. Naturally, such an approach put my back up immediately; of course I'd fight, and told him so in plain terms. More threats came and then when these failed, he tried pleadings, saying that the whole purpose of his visit was to save both me and the legal owner the odium of a court case. But there was still only one answer from me, and though at the time it was disturbing enough, nothing more transpired.

The second event concerned another locomotive, in which an intermediary played a rather shabby part. It was 'Gwynedd', another Hunslet, which someone had purchased two years before from Penrhyn but now wished to sell at £100, and following up a letter from the man who was acting as agent, I telephoned and was told I must make an offer. I was also told it could not, for various reasons, be inspected, and though warned that some repairs were needed and a few parts might be missing, no details could be given. I was assured that in spite of it being 1883 built, it could be made serviceable, and if my offer was high enough it would be accepted. Obviously there was something queer, but just then, with 'The Doll' having to be laid off, I felt we ought to have a second locomotive, and dare not let it go despite the feeling that my arm was being twisted. The intermediary would not disclose the owner's name and address because, he said, the latter did not wish to be bothered until a sale was concluded. The deal was done over the telephone. I offered £150, instead of the £100 originally asked, only to be told that someone else had put in a bid for this amount, but if I cared to go a bit higher, I could have it. It was a maddening situation, yet I felt I must go another £10, and was then asked to put it in writing, with the cheque to be paid before collecting the loco.

When finally it arrived, its condition could have been taken as the reason for inspection not being allowed before purchase. Several parts were missing, there was a hole in the boiler, and the firebox was weak. Having at last been given the owner's name, I wrote, not so much to complain as to get to the bottom of the affair. This brought a prompt telephone call in reply. He said the intermediary had offered to sell it for him, but he had no idea he would use such extortionate methods. He was in fact furious with him, and backed up his apologies to me by returning £25 of the purchase price. How he settled with the intermediary I never knew, but at least the owner was in the clear as far as I was concerned.

18
Fishplates and Dogspikes

The next unpleasant experience of that summer was of a very different type, and its effects lasted a long time. The annual 'Steam Gala' held at the late August Bank Holiday weekend was not expected to draw a vastly increased crowd. This was the fourth and gates had risen steadily each year but one, and this falling off of receipts was accounted for by rainy weather. There were just the usual attractions, with about twenty-four traction engines, Screeton's Roundabouts again, which certainly were a draw in themselves, plus the Cage Bird Show, sideshows and stalls. The gate and parking were manned by relays from the five local organizations, with Adrian in charge. They had worked out a plan in advance using two lanes near the church as extra access and egress from the large meadow, used for parking, since the two little parking meadows were occupied by the 'Gala' itself.

Opening time was at 11 a.m. instead of the usual Sunday time of 1.30 p.m. By noon there was a steady trickle, but by two o'clock cars were coming in nose to tail. By three they were queueing to get in, and the police reported that cars were lined up all the way to Diss in the east, and for at least two miles on the Thetford side of the A.1066 road. Some people were driving on to the grass verge, and were walking in. 'It's beyond us to cope—can you do anything to get them moving in faster?' the sergeant pleaded. With the thirty or so men on parking and money-taking stretched to the limit already, there was nothing I could do. I had my hands full anyway, dashing about as the one in overall charge, but such news was an extra worry, and I could do nothing but hope that the queue would soon grow smaller.

It grew smaller, but not until a good many would-be visitors had given up hope of getting in, and gone elsewhere, or gone home in disgust. When at last we could relax and make some calculations, we found that about 14,000 people had paid to come in (a number had got in without paying), and this was nearly double the previous record. A fine day had brought out far more people than we were prepared for and unexpected traffic bottlenecks had occurred. 'I'll make sure it doesn't happen again,' said Adrian grimly, taking it as rather a reflection on himself as gate and parking organizer. Next day, Bank Holiday Monday, he and his helpers were prepared, but numbers were down. Rain threatened all afternoon, and when it began falling at tea-time, I was rather glad to be in charge of 'Black Prince' placed beside Screeton's Roundabouts in the proper manner, providing lighting from its generator.

People vanished before darkness fell, except for a little crowd of family and helpers in the Bar marquee. It had become something of a joke against me that the only time I drank beer was at the 'Gala', and having downed it I decided to drive 'Black Prince' back on to hard ground, lest the heavy rain now falling made movement impossible. It was nearly dark. Lightning flashed, thunder rolled and the wind increased to a gale. By the time 'Black Prince' was safe for the night, the place was deserted—or so I thought. Indoors the family and some friends and helpers who were staying overnight were clearing up the remnants of the meals Flora had provided as usual for the stewards, judges and other officials. We sat and chatted, counting up takings, and feeling generally content but very weary.

There were ten of us in the house that night, including Anthea and Jenny, who had gone to bed long before. There was also over £3,000 in takings, mostly gate entrance money from both days, which because of the holiday, we had not been able to bank. It was secreted and distributed in various places as we joked about thieves and burglars. The house had never been locked, but though there were risks, we thought it safe enough just to lock doors and fasten such windows as would fasten. At midnight, dead beat, we went to bed.

Flora was the first to wake next morning. She had heard a cat mewing at 6 a.m. and went to put it outside, lest it wake the children. Then she came and woke me. 'The doors and windows are all open—we've been done,' she said hoarsely, with a look of dismay and horror.

It didn't lessen the sense of disgust we felt that the Rally takings had not been found. I fumed to think that anyone could try, as the thieves had tried, to steal money earmarked for charity. It was a horrid thought, anyway, that the downstairs rooms had been gone through systematically, beginning with the kitchen. It was when they found two locked cupboard doors in my study that they really got to work. By drilling round the locks they'd opened them, and in one they found my strong box. This they had carried away, across the garden and a small meadow to a lane. In it were passports, birth certificates, the children's savings certificates and my monthly salary, drawn a few days before in cash, and some spare change. Of course the wild night had favoured them, but if they had been none too pleased to find so little in that strong box, the episode came as a nasty shock to Flora and me, as all this had been perpetrated without disturbing any of us from sleep.

The strong box was never found, and so far as we know the thieves were never caught. We lost much sleep for a long time after that, and home was no longer our castle, until it had been rendered as burglar-proof as we could make it.

On each of the two days, people had queued to take a ride behind 'George Sholto' on its half-mile trip. For most of the time Geoff Sands was driving. Geoff was the son of whom Inspector Sands had spoken so proudly back in 1961, after travelling with me on the footplate of the Britannia No. 70034 to Liverpool Street. A few months later I met Geoff at Crewe, where he was Locomotive Shed Master, though still only in his thirties. This was a treat for me, for at that time steam was only just on the way out in that area. I had to decline the offer of a footplate pass on a 'Duchess' to Carlisle and back, because time did not permit. Instead, there was one to Shrewsbury and back, after an interesting couple of hours spent with Geoff in the loco sheds and repair shops.

By 1966 Geoff had been transferred to the south, as M.P.S. at Wimbledon. As a railwayman he was steadily climbing the promotion ladder from scratch, having begun as an apprentice at the Stratford shops, but as a steam lover he was as happy as could be to drive 'George Sholto' on those two days. His skill went a long way towards demonstrating that a narrow-gauge railway here could pay for itself, and with winter at hand plans took shape for laying a much longer route.

Inquiries for rails had narrowed down to two sources of supply. One of these was at Port Penrhyn, where the Quarry line had ended at the firm's own docks near Bangor. These docks belonged to the Penrhyn Estates and had not been taken over by McAlpines. When on a short autumn holiday in North Wales with Flora and the girls, I'd paced along the various tracks, long since disused, and reckoned there would be a full mile. It was up to me, I was told, to make an offer; but since the ground was needed for development, I had better be quick or the rails would go for scrap.

Back home, I was in a quandary. The other source was at the Claydon Cement Works, only twenty miles away. But news came that after all the track might not be for sale, because it might be needed at another of the Company's works. It would be some weeks before a final decision was taken, and I was advised not to bank on it at all. A great pity, I thought, because apart from the much greater distance, the Penrhyn track would have to be lifted and loaded on transport I must arrange and pay for. And it was not all of one type and size. But I had to have it, and when I found that Roger was agreeable to taking on the job of lifting, with other Bressingham helpers, I came to terms with the sellers, on a tonnage basis.

In early December Roger, Don and Arthur set off to Bangor, and for four days had a gruelling task. Arthur had recently joined us so as to have something to do now that he had retired at sixty-five, having been an engine driver, amongst other things, in his younger days. Roger kept in touch by telephone, and spoke of many unexpected difficulties, not least the rain and gales which plagued such heavy work as lifting rails which had been bedded down for generations. He asked for two more helpers to go. I had miscalculated, and when four lorry loads, about fifty tons, had been taken up and sent on, Roger reported that almost as much remained. But enough was enough, and certainly those on the job had had enough, and were glad to come home. And then came word that the other rails at Claydon Cement Works were after all available.

This put me in another quandary. How, I wondered, could I be sure that the Penrhyn rails would be enough? Could we afford to buy any more—or dare I let the opportunity slip of obtaining a stock against future needs, when lying so close and handy? The Claydon rails were offered at £10 per ton, with sleepers at 1s. each, and with the notion that in the long run we could gain far more than could be lost I decided to have them as well. The plan for 1967 was to lay an outer circle track of about one and a half miles, and with enough track on hand for at least two and a half miles, I chafed with longing to get on with the job as the spring rush of nursery and garden work once more prevented me from doing so until late May 1967.

With another set of points, the outer circle began where the inner line curved towards the pond. The new line ran due south on the west side of the pond, on a down gradient of 1 in 40 till it reached the first nursery field on the black-soiled valley bottom. All this land was soft. In places fine sand lay as sub-soil only a foot or two down; in others it was spongy peat for six or seven feet deep. If you gave a little jump the shudder could be felt by someone standing ten feet away. Such a peaty area lay just where the track had to take a turn from south to east, and to make the formation, a new ditch had to be dug for the whole length of the wide curve.

Ditch digging was no problem. We had an old Massey Ferguson hydraulic digger, and George Orford's skill as driver was equal to most requests we made of him. With George and his tractor to help, I levelled off the formation and bedded the sleepers. A few lorry loads of shingle were dumped at strategic points, and this George brought in the front scoop of his tractor for me to spread. Don and Arthur had let it be known that they would like to lay the rails, and with Don especially having a very security-conscious mind, this was all to the good.

Having rounded the curve on its peaty embankment, the line ran due east for a level five hundred yards. This was easy going since the bordering dyke had been deeply dug out during winter and now the sandy spoil had dried out nicely. I was enjoying myself. The labour was hard but rewarding, and if sometimes I could bed down a hundred sleepers in a day's work, I was very content. But at the end of that five hundred yards' stretch, in the cool shade of trees, there had to be a half curve over a wide ditch, and at that point I had to make a decision. To complete the plan would have taken another four or five weeks. July had come, and if the line was to begin to pay for itself this year, then a short cut must be made. On this low land waterlogging ruled out winter work, and yet it would be in winter that a railway would be most beneficial to the nursery.

The decision to turn north, making a short cut up to the higher firmer land, was a compromise, but easing of the natural gradient —in places more than 1 in 30—posed problems that could only be solved by digging pits for ballast, from the nearest points where sandy subsoil could be found. Once up the slope progress was much quicker, and the link was made with the inner circle at the end of July.

The first locomotive to run on the new circuit was a 1937 Ruston Hornsby diesel. This has been purchased for £10 along with the Claydon Cement Works rail, and needed a tow from a tractor to start its 3-cylinder engine. It had been useful for carting rails and gravel, but I found no pleasure whatever in driving the smelly noisy thing. Three toast-rack trucks had been built on to cement works bogies by a local carpenter, and it was because the new line gave visitors something to see that, even with diesel haulage, some passengers took the new trip.

In having the inner and outer circle my idea was that, with care, each circuit could bear its own train. I imagined that with fares suitably adjusted, visitors would wish to take a trip on both. But it did not work out that way. The first Sunday afternoon's running proved that the longer trip was worth the extra sixpence, and for an hour 'George Sholto' took the inner pond-side route with only a sprinkling of passengers, while the diesel carried far more. Brian Bourne, a volunteer driver who had mastered 'Sholto's' temperamental injectors, was amazed, and so was I. But once we decided to run both trains on the longer route, there was no doubt as to whether people preferred steam to diesel. After that the inner route, pretty as it was, was scarcely used at all, but I longed all the more to have Gwynedd ready for steaming, and have nothing more to do with the diesel for passenger rides.

Brian Bourne had been coming pretty regularly on Sundays during that summer of 1967. He owned a Burrell Traction Engine which each year he brought to the Bressingham 'Gala', he and his family often staying overnight in their living van; though a gas engineer by profession, he possessed real skill with steam. This was not the only reason why I mostly let him take 'George Sholto' round, nor were 'George Sholto' and the little 'Princess' loco in the garden now the only steamers to run on rail. For several Sundays that summer I had driven the metre-gauge Peckett, which we named 'Banshee' and which had arrived the previous December from the Ironstone Quarries near Wellingborough.

I had often stopped to see one of these cross the road when on my way to the B.B.C. at Birmingham, and out of the blue came an invitation to make an offer, since this mineral line was having to close much sooner than expected. As if I hadn't enough on my plate at that time, when so taken up by the acquisition of rails, I went to Finedon, and was told I could have my pick of the three Peckett 0-6-0 Saddle Tanks for £200, so that one at least could be preserved. When a hundred and fifty yards of track were thrown in I could not refuse such a generous offer, having decided already where it could be laid so as to keep the engine alive. A ditch beside the nursery entrance had been filled in a year before with the idea of planting over with something colourful and useful. But this, after all, seemed the ideal place for Banshee which, at twenty-eight tons, was likely to be the largest engine ever to come to Bressingham, and would become a sight worth seeing for visitors using the entrance on Sunday afternoons. Arthur and I laid the track during winter, and by the first open Sunday I was driving it back and forth along its hundred and fifty yards' leash, enjoying it greatly except for the smoke which so often filled the cab. I gave footplate rides, but takings barely covered the cost of coal consumed. The distance was short, and footplate riders too found the cab rather hot and smoky.

19
Gallopers Come Home

Gwynedd, the 'pig in a poke' loco, had stood in the shed for over a year. To begin with, the task of restoration had seemed hopelessly complicated, and it was useless to imagine that once work began it could be continued through to completion. Missing parts had to be made, either by Jack's skill or by sending a pattern to a foundry to be cast. Roger spent several days inside the little copper firebox drilling holes for a patch made necessary by some Penrhyn driver who had used fire irons with too much force. Forty-six patch screws were needed and Jack made every one for less than half the price of new, from a copper rod he had found at home. The engine stood completely dismantled for months, with the boiler raised above the chassis, but though work on it could only be spasmodic it became more urgent as the new Fen track was tried out. The number of people who wanted rides went far beyond my expectations, and even if this occurred only in the space of the two-hour peak period of a Sunday afternoon, that need must be met. It would, moreover, pay to meet it, and to use the diesel as running mate with 'George Sholto' was unthinkable.

Considerations as to whether or not any part of the steam venture brought in revenue had until 1967 scarcely come to mind. This aspect had crept in largely of necessity, but it had formed no part of my thinking when making the garden and allowing the public to see it. As a strictly wholesale-only nurseryman, it had given me a queer kind of pleasure to tell visitors that we had nothing to sell them, and that if they wished to buy plants seen they must approach a retailer. As gate proceeds for charity rose, and as the pressure of visitors to buy increased, it slowly dawned on me that I was being rather stupid not to take some advantage of it. Flora began her stall in 1964 mainly for something to do on Sunday afternoons. Since then sales had increased yearly, and were making some contribution to the firm's revenue; also, a limited retail mail order service for plants unprocurable elsewhere had caught on.

With the steam engines, the process was rather different. There could be no cash return from visitors seeing them on display, and somehow it seemed wrong even to think of it. When the admission fee for adults went up from 1s. 6d. to 2s., I felt first and foremost that visitors should receive good value. The nett amount paid out to charities was still rising, and it was only fair that among the expenses to be deducted from gross gate receipts there should be a payment made to my firm. On the advice of our accountants, this amount was to cover not only wear and tear, such as renewal of worn down grass pathways etc., but coal used by the engines steamed for the benefit of visitors. At £250 for 1966, this sum was well under ten per cent of the total, and was certainly no higher than it should have been. A number of visitors were evidently unaware that nett gate proceeds went to charity, and when occasionally a snide remark about the place being a money spinner for us came to my hearing, it hurt because it was so far from the truth.

It was the addition of a railway that brought in the revenue-earning element from steam. Once the garden railway was completed the potential was plain to see, even though it was limited to a few hours a week for less than half the year. The nursery railway was of course constructed with this in mind, but seen more as a means of recouping the outlay both from the added convenience to the business, and as an added attraction to visitors, who could see colourful nursery fields which otherwise were barred to them. As the summer of 1967 advanced the latter aspect at least was no longer in doubt, and during August a real effort was made to have Gwynedd in service for the annual Steam Gala at the Bank Holiday weekend.

She emerged from the shed only the day before, and still in need of her top coat of paint. Asbestos cement lagging plastered over the steam dome had to make do instead of the correct brass cover which shone so magnificently on 'George Sholto', and it was barely dry as I applied the paint that Saturday morning. Both locomotives were to be in a deep blue, the nearest I could achieve to the old Great Eastern livery. Although Geoff Sands would be coming over again for the Gala as driver for Gwynedd, having given her a trial run, I decided that, apart from these two Gala days, I would appoint myself as her regular driver. For an eighty-four-year-old she ran very well indeed, and even if she was a little off beat, due to having odd valve rods, she had ample power for our needs at the reduced pressure of 90 lb. at which the old Salter safety valves were set. For a couple of hours on each of these two Gala days people were queueing to ride on the nursery line. The trip took only ten minutes, but as one train pulled out with fifty or sixty passengers aboard, the other came back. Passengers alighted on one platform, and then the train drew up to the next, where a fresh lot were waiting.

As usual during these two days I had little chance to drive, having to take spells on the microphone inside the ring where traction engine events were in progress. The gate was down on 1966; the reason for this, we guessed, was that so many people had failed to get in then that they did not attempt to do so in 1967. Even so, about 8,000 people each day in our rather confined space was enough for comfort, and it was good to sense that they were enjoying themselves.

A mishap to the 'Princess' was worrying. It was derailed when carrying a load, and though no trucks came off, the little engine crashed over on its side. Someone with a nasty sense of humour had wedged the spring-loaded points, and then no doubt taken a delight in seeing what happened.

There was noise and bustle everywhere except in the garden, where visitors went to see the flowers and perhaps get away from the noise and smoke. One engine of the twenty-eight or so in steam made no noise, and very little smoke. It was the Tidman Centre on its little trailer in the charge of Bert Swann, and as usual a group of people stood around watching its fascinating motion. Towards the end of the second day I paused to ask Bert if all was well. He smiled, 'Lovely, just lovely,—but for one thing,' he said. 'What's that?' I asked, though I guessed what was coming, since I'd heard it several times before. He raised one of his walking sticks and pointed to where Screeton's Roundabouts were whirling away to the strains of 'The Saints Came Marching In'. 'Them's what's needed,' Bert said with even more feeling. 'Cor—I wish you'd get a set o' Gallopers to go with my little centre engine. I'd be as happy as a sandboy if you'd do that, so I would.'

I laughed and said, 'Maybe one of these days we can run to a set,' though at the time I saw very little prospect. Gallopers were in the same category as steam engines as far as values were concerned. After a period when no one wanted them, those remaining began to rise in price, not only because they had become scarce but because they were virtually antique works of art. This applied to their splendidly ornate organs as well, and as nearly all had been converted to electricity for power and lighting, sets still using a centre steam engine were very rare indeed. The Screeton Brothers were not showmen, but all four were so enthusiastic that they had bought their set for fun, and were spending all their takings, at fairs and such events as ours, on full restoration to its original glory. Each year they came they showed with great pride what had been renewed or restored since last time. And because of such costs, together with the distance they had to travel, they were still charging £100 towards expenses. We liked having them, and each time they left in their caravans they said it was more like a little holiday break for them, compared with the more usual fairground.

In the past I'd dismissed Bert Swann's pleas as impracticable, but with the hectic Gala days over, the thought began to intrigue me. It had been out of the question in the past simply because the money it would take to buy a set of Gallopers was not there. Because the railways were earning revenue my accountant had advised that these and the locomotives should be taken over by the business so as to put matters on a proper footing. The nursery business was becoming more prosperous by 1967, but as it had to stand the cost of a large new potting shed, additional offices, and a house for Adrian, who was the first of my family to marry, there was nothing to spare for a set of Gallopers even if these might be considered as revenue earning. Robert too was planning marriage and would need a house, and the business would have to foot the bill for a largish shed I hoped could be built in which to store the steam collection.

Then suddenly a way opened out when I was studying one or two advertisements in *World's Fair*, one of which was for a set of Gallopers, with organ but no centre engine, at a much lower price than I'd imagined would be asked. Just at the same time a visitor had called and asked if I would sell the Marshall. At first I bluntly refused. But when he backed his request with a very good offer, saying he had sentimental reasons for wanting this particular model (which, being a double-crank compound, was rare) I said I'd consider it. This engine had never been a success and Jack had been baffled all along by its proneness to knock somewhere on the crankshaft. It was in fact the least popular of all the engines though it looked so smart, and its turn to be steamed on Sundays by one of the volunteer drivers was often skipped. 'Let it go,' Jack advised, and so said Roger and Arthur.

I told the man about its fault, but he was in no way deterred. 'I want it', he explained, 'because my father owned one just like it when I was a boy. I've just retired from being an engineer by trade and have made a model of it. This has whetted my appetite to own the real thing, and I'm prepared to buy if only you'll sell.'

I let it go, and remembering my Endowment Policy which had matured the year before and which I had lent to the business because of the Bank Credit Squeeze, I decided to follow up the advertisement in *World's Fair*. 'Would your own set of Gallopers appeal to you?' I asked Flora. Her eyes lit up quizzically. 'You know very well there's nothing I'd like more,' she said—'but what's the idea—you can't afford a set, so you must be teasing.'

I assured her I was not and then told her what was in my mind. It had occurred often enough that over the years my avidity for collecting engines had been to some extent at her expense. There had been times when I'd put engines first, either through skimping improvements inside the home or through being too preoccupied with restoration work to take a break with her away from home. Now, with money matters improving, there might be a good chance of expressing appreciation in a tangible way by making her a present which would be in keeping with the steam collection as a whole. If I got a set of old-time roundabouts, and made over the Tidman centre engine as well, it would be complete. And Bert Swann, too, would have his dearest wish fulfilled.

Roger was all in favour of such an acquisition. Though so dedicated to steam, he had in earlier years had leanings towards showground equipment, being especially thrilled by fair organs and the less modern rides such as Gallopers. He had also considerable musical gifts of his own and apart from often playing an electronic organ at dances, was a regular church organist in his native parish of Garboldisham. It was in fact his copy of *World's Fair* which he had brought over in order to show us the advertisement for a set of Gallopers.

Early one morning, Flora, Roger and I set off for Scotland, catching an Edinburgh train at Peterborough. But it was growing dark by the time we finally reached a small mining town in Fifeshire where the Gallopers were still erected. It was one of those hurried deals—not only was time short, but also I had more or less made up my mind in advance. The owners were fairground folk who had, I was soon able to discover, even more firmly made up their minds to sell, saying they believed some more modern type of ride would pay them better. They reckoned too that modern canned music was more popular than music from an organ, and because of this theirs was seldom used. All it needed, they said, was tuning; knowing no better, we believed them. When it was played for our benefit some almost unrecognizable tunes emerged until we asked them to switch off the blaring, jarring noises it made. But because they were willing to come down well below the advertised price, and deliver the set to Bressingham, and because they seemed pretty genuine people, we clinched the deal within an hour, over a high tea they provided for us.

It was dawn again when we left the mail train at Peterborough after a sleepless night because the only passenger coach was so full. For a week we could scarcely believe the reality of what we'd done. But there was no doubt about it when two large ancient lorries arrived, each with its trailer. One lorry, with generator underslung, was left behind after the other had disgorged its thirty-six horses, ostriches and cockerels, sadly tarnished and chipped but still gaudy. And left in the yard under its tarpaulin stood the centre truck, bringing back in a different context memories of a sight which had struck terror in my early childhood. It still looked like a gun carriage, with the centre pole folded down just like the barrel of a cannon, grimly menacing.

Flora was quick to find out more about its history. The makers were Savages of King's Lynn, who were still functioning as an engineering firm. She learned that the horses and birds had been hand carved, probably before it was made into a steam-driven set in the 1890's. The organ's history was more difficult to trace, but most likely it was made in Germany about the same period, judging by the dress of the carved little figures, poised in front as conductor and shepherdesses to tinkle little bells. Thrilled to bits, Flora could scarcely wait to make a start on the long process of restoration, beginning with the horses. 'It will take me years,' she said, 'but it's got to be right, and I want to do all the repairing myself.'

20

New Hopes, New Problems

That hurried journey to Scotland had followed another journey north made not long before. It was a journey purely for pleasure with nothing in prospect except to be on a train hauled by the 'Flying Scotsman' on the 'Cathedrals Express' to York and back. And pleasure it was, in spite of my wishing that a rumour in circulation the day before could have been true. The Norwich B.B.C. telephoned to say they had it on good authority that I'd bought the 'Flying Scotsman' from Alan Pegler, and was I taking it in after this trip to York? I was not in the house and Flora had taken the call. It struck her as so amusing that she played up a little, replying that although I was liable to spring surprises on her, she couldn't imagine how I could have kept this one quiet. She knew well enough it was no more than a rumour, and finally said so.

It caused an introduction as well as a laugh on the journey with Alan Pegler, but the rumour had been passed on to the ears of a senior B.R. official, also on the train, with whom I had corresponded but whom I had never met. Inevitably, as we chatted, I said how much I regretted the fact that a full-sized B.R. locomotive was out of reach and told him of my visits to Doncaster and Colwick loco depots three years before, which had come to a dead end. 'Have you seen what it says in the new Transport Bill about preserved B.R. locomotives?' he asked, in a way that made me wonder what implications his question held. I admitted I'd not studied the Bill, beyond taking in some of the pretty drastic features recently brought to light through the newspapers. 'I'm interested in preservation,' he went on, 'and in the new Transport Bill it states, among other things, that Clapham Museum will have to close, but what may interest you is that the British Railways Board may now sanction the placing of locos scheduled for preservation in public or private museums if their requirements can be met. Your collection at Bressingham, I'd say, is more comprehensive already than some museums, and I think it would be worth a try—that is, if you think it's worth while.'

He must have seen the almost incredulous look on my face, because I caught just a glimpse of my dearest pie in the sky wish taking shape. 'It's like this,' he continued. 'There are locomotives on the preserved list which are still not allocated to museums. They're in store and a constant headache because of souvenir hunters and vandals. You've already shown what you can do with old traction engines at Bressingham, you have lots of people who come to see them and your gardens, there's nowhere else in East Anglia where there could be a B.R. loco on show like that. You're a good customer of British Railways, and we in Norwich will back you up, if you can put your case to the Curator of Historical Relics. I'm pretty sure you can, and I think you might stand a reasonable chance of getting one on permanent loan.'

It was as well that this conversation took place after the return journey began, or I would have been too distracted to take in what was to be seen at the York Railway Museum, or to enjoy the thrill of the journey up. With renewed hope of having my fondest wish fulfilled, I had to restrain my imagination from running wild and throttle down to the one avenue, now that a possible entrance to it had been revealed. It might be tortuous, it would take time, and it might lead once more to a dead end, but if it did it would not be for want of effort on my part. I'd been told in so many words that if my application was successful a B.R. loco would cost nothing, at least up to Diss station. What happened after that would be a fence to climb when it was reached. A way would surely be found if that time came, and the more I thought about a locomotive on permanent loan, the better I liked the idea. Personal ownership of engines was already losing its appeal for me, since my collection had now become what was probably one of the most comprehensive in the country. Values had risen, and I was becoming increasingly concerned to ensure, as far as I was able, that it should remain intact beyond my lifetime, as a collection of public interest. I had in fact already made tentative inquiries on this point with both my solicitor and accountant, thinking that some form of Trust was the right solution. My sixtieth birthday had recently passed, and was a reminder of inevitable mortality despite the energies I was still fortunate enough to possess in good measure.

This new prospect called for mental energy. Physical energy was having its outlet in laying an extension to the Fen railway. The short cut back up the slope had to be lifted after less than a season's use. Judging from the preference visitors had for the longest route open to them, the loop planned but not completed early last summer would be the regular run, and I decided to concentrate on it now, having carefully mapped it out. One variation of the original plan was to bear north-east from the set of points where the loop began. My first intention was to go due east as far as our own land extended, but Robert intervened. 'Have a heart,' he said feelingly, 'if you cut straight across Roydon Fen, I've got to fence it in against cattle, and they'll need two level crossings. Can't you just cut the corner off, so as not to carve up the grazing into four little bits?'

I took his point. It was very poor grazing pasture, having been reclaimed from a swampy waste from which years ago turf fuel had been dug out so that the land became permanently waterlogged, too low for cropping and often flooding in winter. So it was that a curving ditch was dug for the soil to serve as an embankment on which to lay the track. Beyond this, the curve went back from a north-east to south-east direction, as it gained slightly higher ground. This area went by the name of Freezen Hills and until we tackled it in 1957 it had been a waste of another kind. It was a low mound of sand which for generations supported nothing but rabbits, gorse and brambles, apart from silver birch trees and a few small oaks. The County Agricultural Authority had failed to trace the owner when wishing to serve notice to him of their intention to destroy the rabbits, and suggested I and a neighbour register legal claim to it. This was done, and though we cleared the bushes, levelled off some old sand pits and seeded it to pasture, the higher sandier parts of this twelve acres were still too poor and hungry ever to recover the outlay.

But this year, after a trial patch under certain kinds of plants, it was all being ploughed again. There were kinds that had proved to have no objection to this acid sandy soil, and because it was to come for nursery in 1968, I decided to encircle it with the railway first, to give a total run of just over one and a half miles. For me, levelling the sandy soil and laying sleepers, it was again enjoyable work. I was glad we'd left a score or more of oaks and birches as windbreaks, for now the railway could curve effectively between them. And when visitors took to the train, they should appreciate a colourful pattern from the acres of phlox to be planted there along with red hot pokers and a variety of other plants, as they emerged on the return alongside yet more nursery fields they could not otherwise see. With summer gone, the railway could be used to bring back loads of plants for despatch, since it served this distant and least accessible portion of the nursery.

Fred Orford, the neighbour who jointly staked a claim with me on Freezen Hills, gave me permission to cut off a corner of the piece of land on which he ran poultry and pigs. The rest of his share was already added to the nursery for convenience' sake, in exchange for one of our stiffer-soiled upland fields by mutual agreement. This corner piece enabled me to lay the formation on a much easier curve and gradient, so as to link up with last year's track at the top of the slope where the short cut had ended. That short cut had cost me a fiver the first time 'George Sholto' climbed it. In opening the regulator to take the 1 in 36 bank, a live spark had settled on a lady's lap and ruined her dress. After that, whoever was driving had to make sure of a full head of steam and take a run at it from the bottom, so as to avoid more claims, even though it was exciting to make the engine bark.

Don and Arthur, at odd times during the winter, resumed the job of laying the rails. With rails spiked down to the sleepers, the diesel followed up, coughing away all day long because of not being able to start the engine without a tractor towing it off. When the circuit was completed, the diesel brought along truck loads of gravel and the laborious task of tamping with bars and shovels to underpack the sleepers had to be done along the whole route. Settlements, especially over the soft peaty places, had occurred with usage and must be corrected. But when Don said a job was done one knew it was well done, even if he let it be well known just how much was involved in the doing. Some of the rails had been damaged before leaving the cement works, and kinks could only be straightened with the 'Jim Crow'. This was a clamp designed for straightening or for making a curve in flat-bottomed rail, and I've never discovered the origin of this curious name.

It was a relief, for more reasons than one, to have the railway finished before spring brought its exacting demands. The New Year held every prospect of being hectic all the way through as far as railways and locomotives were concerned, and by way of relief from heavy manual labour I repainted the Beckton 25 during a mild spell in February. This little shunter had in fact arrived during the previous summer, as the first standard gauge loco to come to Bressingham. I had become a member of several steam preservation societies over the years, including the Industrial Locomotive Society. The Secretary of this small but effective Society, J. B. Latham, had in the course of correspondence told me that a new home was wanted for this Neilson 0-4-0 presented to the Society by the North Thames Gas Board. He wondered if I would be interested, and when I replied that I most certainly was, it came in due course to stand for the time being beside the metre gauge Peckett. It was a curious little thing, built in 1897 and standing only about five feet high though it weighed twenty-two tons. Someone dubbed it 'the beetle', and indeed this was rather apt, with its saddle tank looking like a shell. It needed a scrape down and repaint, but apart from this was in quite sound condition. The low build had enabled it to be driven under the gas works retorts, so I was told, but it was not until some months after its arrival that I had more than a vague idea as to what to do with it.

Vague ideas have a habit of taking on clearer outlines if one hangs on to them. What had existed nebulously in my mind at the beginning of 1967 had become a subject of paramount importance by the end of the year. It was the need for a building in which steam engines could be housed and cared for; a place specially designed and built for them, but two major problems, apart from cost, had first to be overcome. The first was the problem of finding a site, but linked as it all was to size and cost, there loomed at the back of the whole concept the possibility of obtaining at least one British Railways locomotive. This implication, and the imaginings the prospect brought, had opened up a new dimension, because it could so easily lead on to a development beyond my wildest dreams—a full-scale live steam museum.

The first site to be considered for a museum building was the field on the east side of the nursery entrance. For years this had been under nursery cultivation, and as is always the case with hardy plants, it had to be replanted every year. It was the handiest field for the packing shed, but handiest also, I recently decided, as additional parking space for visitors. The previous year had proved that more would be needed, and it was being left vacant for 1968 so as to sow it down to grass when spring came. Trying to imagine what appearance a building placed there would have, I finally ruled it out, as being too close to the main road. The only other site available presented the problem of levels. This was on the south side of the broad yard between the farm buildings and nursery office, and a check revealed a slope of 1 in 30.

This was where the overriding problem arose. If designed to house rail locomotives as well as road engines, and if it was to be a live steam museum, then the floor of the shed must be raised to a level which would ease off the slope up the yard, on which rails might have to be laid. For the last three months of 1967, this had been an anxiety from which there was no escape until word came as to whether or not a B.R. locomotive would be allocated. Of one thing, however, I could be certain, and this was that no allocation would be made unless the building was adequate in every way. Until October I was thinking of a shed that could be built in at least two stages so as to limit the capital outlay to existing needs, by adding on only if and when needed and available capital increased. But with the possibility of a B.R. loco looming, it soon became obvious that a plunge had to be taken in order to have an adequate building as backing to my application.

Once I had made the decision, Robert and I between us planned it out. The site allowed for a hundred feet in depth, and though the width was less at that time because of a concreted compound in which sugar beet and pig manure were heaped in season, some of this would have to be taken in, to give roof coverage of a hundred feet each way. For cheapness, as well as durability, steel stanchions and trusses would form the skeleton, and these could be made locally. Asbestos roof and concrete block walls could be erected with little or no outside help, but the first problem to be overcome was the slope of three feet or more from front to back. It so happened that a nearby farmer had recently excavated a large amount of subsoil for a new cowshed, and since it contained some gravel in the mixture, we believed at the time that it would serve as filling material. When we first considered it, it was reasonably friable; when it was carted in—some 200 tons of it—heavy rains came. Once there on the site it had to be spread with a bulldozer, which in the process of spreading turned it into a sticky, dough-like mud. And knowing that was how it would remain until spring came, or until it was roofed over, the prospect of erecting steelwork and roof was not inviting. A severe frost would naturally harden it, but this in turn would bring its own difficulties and discomforts.

But as we had ordered the steelwork the job had to go ahead. Levels and angles had to be accurate when digging holes to fill with concrete for the supporting stanchions. Drains had to be dug before the foundation wall footings could begin, since rain water both from the sloping yard and the roof must be conducted away from the shed. What with one thing and another—from the over-riding uncertainties as to what would eventually occupy the shed, to those connected with the actual work, depending as it did on planning permission—the end of 1967 found me with rather more anxieties on my mind that I'd had for many a year. When trying to sort some of them out, invariably the same conclusion was reached, that all this trouble was of my own making and I must therefore carry the can. Some may have thought I was pushing my enthusiasm for steam preservation too far—that I was fast becoming an eccentric, if not a fanatic. My gift to Robert of the Sentinel Steam Tractor as a wedding present that Christmas might well have confirmed this view for some of the guests.

21

Hopes and Hazards

The great news came first by telephone from Captain Peter Manisty, Chairman of the Association of Railway Preservation Societies. He knew how anxiously I'd been waiting, and had played an important role in the complicated negotiations with British Railways Board. Once I'd written my letter of application, I had no idea of the procedures that would follow. This was where Peter Manisty came in. As Chairman of the A.R.P.S. he was also a member of the Consultative Panel, and able to make recommendations to the B.R. Board on the allocation of the historical relics which Mr. Scholes, as Curator, had under his charge for disposal. I had been invited to a little informal meeting at the Clapham Museum in November, with these two and Mr. R. C. Riley, another member of the Panel, and was shown a list of rolling stock scheduled for preservation but still not allocated to a museum. As I studied the list, almost boggle-eyed, I blinked even more when asked which of the dozen or so I would like, scarcely believing I would have much choice.

The names that stood out for me on that list were 'Britannia' and 'Thundersley'. When I came out with them Mr. Riley spoke up to say that 'Britannia' was in a deplorable condition, having been badly pillaged before being shedded at Brighton for greater safety. But Mr. Scholes said, 'Would you really give a home to "Thundersley"?'—from which I could only deduce that this 4-4-2 tank was languishing. My thoughts went back to the little model built by Commander Robinson, which I so heartily wished I'd never parted with. At the time, two years before, I'd been hard up enough to fall for an offer for it at a modest profit, and knowing I would never use it in steam, had let it go to a man who, I discovered too late, was a dealer in spite of his assurances to the contrary. And now there I was, it seemed, being offered the original 70-ton 'Thundersley' on which the $1\frac{1}{2}$-cwt. model was based.

'I'd give it a good home,' I said feelingly. It was then that I learned that the final decision rested with the B.R. Board, and must wait until the Consultative Panel had met and made its recommendations to the Board. We chatted, without my realizing at the time that I was to some extent being vetted in the process, and for good reason since I learned also that mine was the first application of a private nature. These three gentlemen were fully in the know about every aspect of locomotive and rolling stock preservation, and as I knew so little but wanted to know more, I listened very intently to them. The discussion went back to 'Britannia'. As the first of the B.R. Class 7's to emerge, it was the one designated for preservation under the guiding rule. It was a pity, they agreed, that such a rule applied in this case, because No. 70000 was so rough, when others of the class were in much better shape, especially 70013 which had been overhauled at Crewe only a few months before—the last steam locomotive to go through the shops. A cloud of uncertainty came over this conversation. Rumours were afoot that steam was to come to an end before 1968 was out, and though it was known that No. 70013 was to be used on special steam-hauled trips until then, rather than on regular service, it would in the prescribed course of events be sold for scrap after that unless someone could purchase it outright for preservation. A Mr. Cheetham was mentioned, whose role under B.R., it seemed to me, was rather like that of Thomas Cromwell under Henry VIII. It would, I gathered, be very difficult to persuade Mr. Cheetham that No. 70013 'Oliver Cromwell' should be saved by the preservation order, and 'Britannia' should carry the death sentence instead. And there the matter had to rest because it was time to go to the Annual General Meeting of the Transport Trust, which all four of us were to attend.

A few weeks later, with a couple of hours in London to spare, I took a chance on finding Mr. Scholes available at Clapham. Under his Curatorship the Transport Museum there had become a fascinating place, but he had a vast amount of other work besides, with every conceivable item connected with railway history his responsibility. I was shown into a large building where huge stacks of documents, plans and equipment were stored, and among a variety of station boards one taken from a station near Leeds was fixed above an inner door. That name was 'Scholes' and through this door Mr. John Scholes came out to greet me, inviting me in for a chat. At first I began to wonder if my application for 'Thundersley' had made any progress at all. But before long I learned that a strong recommendation in my favour had just been received from the Divisional Manager's Office in Norwich, and it was quite evident that this was the kind of backing Mr. Scholes needed to go ahead. I could see now how vital it was for him to be quite sure of my ability to accept and discharge such a responsibility. I had nothing specific to ask him, but described in greater detail what my plans were for a live steam museum, mentioning that the new building was to give coverage to three standard-gauge tracks in the centre, with narrow gauge on one side and road engines on the other. 'Then you'll have room for more than one B.R. locomotive?' he suggested. I nodded hopefully and asked by way of a lead if he still thought it worth a try to get the preservation order switched from '70000' to '70013'. He was quick to take my point. 'Aha,' he laughed, 'so you'd like to see "Oliver Cromwell" come back to East Anglia?' 'And the first Oliver Cromwell was an East Anglian too,' I put in, adding that apart from the Britannia class being my favourite locomotive, it would be rather appropriate since many had served to give a real fillip to East Anglian lines. Mr. Scholes agreed heartily, and after explaining again the difficulty personified in Mr. Cheetham, finally made up his mind to have a try. 'I reckon there's a good case,' he said with conviction. 'Leave it to me, and I'll have a go tomorrow.'

What Captain Manisty told me over the telephone three weeks later was by way of a tip in advance of official information. Not only was 'Thundersley' to be allocated to Bressingham, and moved into Norfolk in readiness without delay, but the Board had also recommended that when 'Oliver Cromwell' was withdrawn from service, this too should come. When that day would arrive, no one seemed to know, but for the present the fact that it was coming must be a secret. I could scarcely believe my luck, but it was officially verified almost immediately by the Norwich Divisional Manager, and it touched me pretty deeply to realize how much sheer goodwill had been at work in my favour while I'd been chafing with impatience during the past weeks.

The closer one gets to holding in one's grasp something long hoped for, schemed for and worked for, the harder it is, I've found, to remain patient. When there is nothing to do but just wait for final fulfilment, not knowing how long the waiting will be, then it's very trying indeed. 'Thundersley' had been promised, and that was fine, because at the outset I'd never allowed my hopes to go beyond one B.R. locomotive. In my imagination that one alone would be enough to satisfy my 'live steam museum' ambition. Now, with hopes alive and kicking that 'Oliver Cromwell', as a representative of the class which for me was the zenith in British locomotive design, was also headed this way, I was so disturbed that I could think of little else.

A journey to Yorkshire in company with Norwich B.R. officials helped to keep me down to earth and to the present. There at Hellifield, on a bleak hillside, we were shown into a large locomotive shed. Its leafy roof dripped melting snow on to the few locomotives there in store, and one of them was 'Thundersley', a long way from its original London, Tilbury & Southend tracks for which it had been built in 1909. It stood there, handsome still, but forlorn and neglected, so obviously in need of a good home. As we took a closer look, inside and out, I learned how it had been restored and painted in its original green livery for the L.T.S.R. centenary in 1956, but had scarcely been used since. Like many others scheduled for preservation it had been plundered when stored at Stratford, and had come to Hellifield for greater security. Even there it was in danger, and we saw how some vandal had tried to force apart the fine L.T.S.R. crest from the side tanks with a tommy-bar. This had failed, but the marks were indelibly there.

Because the Hellifield shed was not fully secure (thieves had climbed down from a smoke vent in the roof), 'Thundersley' was under orders to go to Preston Park at Brighton to join other locos in store, but now that transfer was cancelled and the next step was to get her to Norfolk. At this point a suggestion came up from members of the Norfolk Railway Society. The new shed at Bressingham would not be ready for several weeks, but the need to move it away from Hellifield to some safer spot was urgent. A lineside goods shed at Attleborough could be made available, and with a signalman on duty nearby, it would be safe. Once there, a team from the Norfolk Railway Society would undertake to remedy at weekends the ravages of time and neglect, as far as they were able. One of their members was Mr. D. W. Harvey, a steam engineer of high repute, and he would be willing to replace the motion and attend to such fittings as the steam brake and reversing gear, which among others were seized up with rust.

This was a splendid offer which I gladly accepted, though none of us realized how long it would take. Until the transfer of 'Thundersley' took place in March I had little idea of how complicated a matter it was to tow a locomotive over a distance, but was thankful it was not I who had to make the necessary arrangements and see that they were carried out. It was only twelve miles by road to Attleborough, a small town on the Ely–Norwich line, and a little party of us went from Bressingham to see 'Thundersley' arrive on a March day so windy that I was glad of the excuse to leave my task in the garden. In any case I would not have missed the sight, pathetic though it was to see a noble old steamer hauled in behind a diesel. But unlike some I'd seen being towed to a Norwich scrap yard to be destroyed, this was to come to life again and stay that way.

The windy weather had come too late to be of help in drying up either the sodden nursery and farm land or the site of the new steam shed. It had been a wet but mild winter till March, and on all sides work was behind schedule. It had taken a long time to erect the steel work and lay the concrete footings of the shed. Holes and trenches were dug through the sticky clay mixture only to become half filled with water overnight. Beneath the filled-in overlay the soil was sandy and I had considerable misgivings about consolidation for the weights it would have to carry. But we had to press on regardless, as far as the weather would allow. It was a common occurrence for one of our helpers working on the site to leave a gum boot behind in the sticky mud, and there were times when I was so frustrated for lack of progress that I made little runnels with a spade in an attempt to run the surface water off.

At last the time came when, with steel stanchions erected, the roof trusses, also of steel, could be lifted on. This called for a crane, and although we had left the erecting for a time in the hope of drying out, in the end we had to bring in fifty yards of crushed airfield concrete to make a solid roadway on which a crane could operate. Crane hire costs are high, but a slight error on the part of the man who welded the trusses caused several extra hours to be charged. With trusses in position our own helpers, under the intrepid all-rounder Graham who was normally an expert packer of plants, scrambled up to fix in the purlins. It was twenty-eight feet to the apex and at that juncture March winds began to blow, making the task quite hazardous. A load of asbestos and perspextype sheets for the roof arrived just before the first gale, and someone omitted to weight them down. In the morning most of the roof light sheets were scattered and broken, and a new portable nursery shelter was carried two fields away and smashed to smithereens.

It was all very trying. More gales blew up when the time came to place the roofing sheets. One was so fierce that work had to stop, despite the sheltered position of the building. The sky took on a peculiar colour from the dust blowing off the Fens and the sandy lands to the west, the direction from which spring gales can cause havoc, as well I knew from bitter experience of twenty-five years ago, when as a farmer in these bleak Fens I'd known dykes to become filled not only with dust, but also with the seed and fertilizer. I had become a little more philosophical since then, but my propensity for biting off more than I could comfortably chew was, it seemed, undiminished over the years.

There was no time now for musing, and no sense either in pausing to wonder why so often some fairly big project I'd undertaken turned out to be much more toughly formidable when I was at grips with it than when I had first conceived it. Once again the job was calling for far more effort than expected, once again my optimism was shown to be at fault.

The news that 'Oliver Cromwell' was allocated to Bressingham had been decidedly cheering. It put fresh heart into the few who could be told, but the very fact of the injunctions to keep it a close secret was in itself more than a little disturbing. Few reasons for this secrecy were given, and there seemed to be an implication that with No. 70013 some doubt still existed. If its destination became widely known, then someone with influence could throw a spanner in the works with the argument that such a locomotive, famous already and likely to be more so as the last operational steamer to be run by B.R., should be retired to a national museum and not to an obscure village in the heart of East Anglia. I gathered that there was to be a special last steam run during the summer, to mark the closing date of the railways' steam era. The date was not yet fixed, nor was anyone I knew aware of the route this swan-song run would take. But I was warned by our official sponsors, who had worked so effectively in the high places of B.R., to keep it quiet. I was not going to let them down, risking perhaps at the same time an almost unbearable disappointment for myself.

I could at least confide in my family. Flora was still hoping for improvements within the house. Adrian had been hoping 1968 would bring the much-needed new packing shed, and though Robert would have liked a new piggery to be built, he was for the time occupied with his own new house on which work was about to begin. All three had been pretty tolerant of my overriding decision to build the new shed. In a way, it had been an act of faith as far as the big B.R. locos were concerned, and now that 'Oliver Cromwell' was earmarked, up went my sights once more.

As it stood, an area of 105′ × 100′ would be covered. The west wall had been left till last, largely because of drainage difficulties, and now I was glad, because the headroom allowed for a further extension of 18 feet, costing very little more because the outer wall would not need to be so high. 'Some shed!' was often the remark of occasional visitors that spring. So it was extended to 123′ × 100′, but I also knew that before the great day came when 'Thundersley' arrived, as the first B.R. loco on permanent loan, Mr. Scholes would be inspecting our premises and it was his opinion that would count. He had only my word, so far, that these would be up to the required standard in area and stability, and he would want to see as well what precautions we had taken to make the floor safe for the weight it had to carry, to say nothing of precautions against theft, which had so bedevilled his job as curator of historical relics.

In for a penny, in for a pound. Now that the project had got so far, it would have to be carried all the way. I had an uncomfortable feeling that the shed would finally cost more than the first estimate we'd made, but at this stage I didn't want to know by how much. It would set the firm's resources back appreciably, and once again other, and strictly more vital capital investments would have to be postponed. But to me this one had become vital, and I had no qualms about the future—provided that no hitch prevented 'Oliver Cromwell' from becoming the guest of honour in that new shed.

A visit to Kingmoor sheds at Carlisle was arranged. Steam there had just come to an end. Following the withdrawal from service of the last Britannias which had been based there, many were standing on sidings in dismal rows. Most of the name plates had vanished, and they were pitifully forlorn and neglected, as were several of the 8 F class as well. It was a sad and depressing sight, and regrets were expressed by those who showed us what was left in the stores. It was difficult to know what might be needed at some future date with 70013 at Bressingham, but a selection was made mainly by Mr. Harvey, the steam engineer in charge of 'Thundersley', who was one of the little party. It was the last chance, for soon all was to go for scrap at so much per ton, and I could not possibly miss this opportunity for securing spares.

22

The Three-piece Loco

At almost every stage of the steam venture it had been a case of feeling the way as opportunities for development presented themselves. This applied every bit as much to public openings as to the building up of the collection of engines, and the more the collection grew, the more people it attracted. During the summer open season it had become increasingly difficult to ward off visitors who either did not know when we were open, or tried to disregard the fact that we were open only on Thursday and Sunday afternoons. A notice board at the entrance stating that the nursery was not open for retail business or to casual visitors may have deterred some, but others still came in, asking to look round the gardens or to see the engines. They were a worry, and if any suspicion arose when accosted that they were trying it on, out they had to go. We had a business to run and we had to draw the line. Occasionally, when someone had made a genuine mistake and had come a long distance, then we gave way, but by 1968 the advertising was designed as much to prevent people from coming when we were not open, as to attract them when we were. Those who came expecting to buy plants to take away put my back up most of all, and if they got uppity, as some did, I quickly grew shirty, especially if it were a Saturday afternoon when the family and I were on our own. If their interest was in steam it was different, for mostly they had made a genuine mistake and were reasonable people.

Over the years the opening period had become approximately fixed. Pressure of work on the horticultural side did not ease until late May, but began again in autumn. This was also the period when our grass parking area could best stand the wear and tear. One experimental Easter opening had played havoc with the grass both in the garden and car park, since it followed heavy rains. In 1968 the last Sunday in May happened to be Whit Sunday, and for the first time we conflicted with the Woodton Rally, because we had to decide on opening on Whit Monday as well. Once before when the open season began before Whitsuntide it had been like it used to be with August Bank Holiday Monday, a thankless task for someone to keep people out. Our being open at Bressingham was not the only reason why, for the first time, none of our engines was on show at Woodton. Apart from the need for our visitors to see the engines, our drivers were needed as well. It nagged none the less, until I learned that in spite of the 4,000 people who came to Bressingham each day on Whit Sunday and Monday, it made little or no difference to the usual receipts at Woodton, twenty-five miles away. Not that I really believed it would, because what Bressingham offered in variety was so different from the concentrated pageant of traction engines offered by the Norfolk Club at Woodton.

I was sorry to miss this event, but neither I nor my engines could be in two places at once. Since 1962 I had been a Committee member and Vice-Chairman of the Club, and had enjoyed the dinner and social evening always held in Norwich during October. For several years we had organized an informal evening in steam at Bressingham, during January or February, mainly for Club Members, which helped bridge the dull winter months. With the engines packed in under cover, there was not much room to spare, and if the air was smoky, there was the warmth and smell of three or four engines in steam to enjoy, with hot refreshments available. This little gathering had to stop after 1967, with so much on hand, and so little indoor space left; and with the need for a holiday to be met only by going when the open season had ended, we had to miss the Club's October dinner and Annual General Meeting as well.

It was the variety, it seemed, that attracted visitors to Bressingham in increasing numbers. The whole thing was snowballing, and not the least important new feature for 1968 was Flora's Gallopers. She had spent all available time for months past in repainting the horses, ostriches and cockerels which now glinted so splendidly in the light. Most of them were horses, some with intricately carved manes, and with Adrian's wife Rosemary to help, one by one they had had their gay patterns lovingly traced and coloured with flamboyant paint, and were stored up in a spare bedroom till all thirty were ready. I was doing a series of Gardening Spots in Anglia's 'At Home' programme during that time. The TV Unit came one day for this, but as soon as they saw Flora and Rosemary at work the producer changed his mind, and my gardening demonstration had to take second place.

We built up the Gallopers just south of the office, where for years a little patch of ground had been used for heeling in plants. Now it was sown to grass, and the centre truck with its grimly cannon-like folded chimney was placed in position. Around this, till its chassis was completely hidden, the ancient set was built up, with the organ facing one way and the Tidman Centre engine the other. Bert Swann came to inspect and boards were laid to form gently sloping gang planks so that he could shuffle up on those legs of his that would not bend. The perch beside his engine was less handy now than before, but because the Tidman was now to serve the purpose for which it was made he was highly delighted. On the rounding board, instead of the former high-falutin' showman's advertisement was the legend—'The Bressingham Steam Roundabouts, Built by Norfolk Craftsmen when Victoria was Queen'. And when it began to revolve, with mirrors flashing, with the bright paintwork glinting under the lights, and the organ playing, it was very attractive. Even so, it was annoying when now and then someone referred to it as a 'hurdy-gurdy' or asked if the next thing would be a coconut-shy or a hoop-la. To us, as to most people, it was an antique of a very nostalgic and colourful type, not out of place because of its steam element. It would give innocent pleasure to lots of people and if what they paid for rides would enable Flora over the years to restore it fully, then she too would be content.

There were three narrow-gauge locomotives in steam that Whitsun weekend. 'George Sholto' had at last been painted in Great Eastern blue and was hauling three new toast-rack coaches built by a local carpenter on to cement works bogies. 'Gwynedd', too, was in service, with me as driver, and for a time passengers flocked in to make the rather hard-seated trip of one and a half miles round the newly extended line. The third loco in steam was 'Maid Marian', but its attendants were beginners and found working pressure difficult to achieve. They were members of the 'Maid Marion Fund', a group of enthusiasts who had subscribed to purchase this 1903 Hunslet from the Dinorwic Quarries in North Wales.

Having made the purchase, they had no track on which to give expression to their enthusiasm, and had made a tour, some months before, of all possible tracks of their gauge. There were only three or four in the country, including Bressingham, and having seen the others they accepted my offer of a home and facilities for running once a month. It could scarcely be oftener than this because nearly all the members lived in Lancashire. And though for the first time they were indulging their love of steam, they were not very happy. Their 'Maid Marian' had been restored, but as always there were little snags—injectors that wouldn't work, priming, leaking glands and the like, but by the end of the day with both fire and water in full harmony, having tried, erred and finally mastered the mysteries of steam, they were able to chuff round the track to their hearts' content.

Since the advent of railways and locomotives my interest in steam had broadened, but it was no case of switching from road to rail. Bertha, the first to come in 1961, had been re-tubed and repainted after being stripped down, and Cissie the Fowler Ploughing Engine purchased from Mr. Stebbings was putting her sister Bessie to shame, in her new garb of royal blue which Robert had lined out skilfully in red. Then there was the Burrell single-cylinder Traction which had come from North Norfolk. It had stood there out of sight in an estate woodyard and had been used for driving a saw for twenty years before being offered to me. It was the nearest now to Bella, the single-cylindered Burrell I bought in 1947 and which had come to such a sad, untimely end just before my return from Canada. This too had been restored and was in its original maroon livery with yellow lining. Then there was the Youngs' Portable, which Jack greeted on arrival as if it were an old friend. He had been a foreman fitter for Elijah Youngs who until 1940 owned a foundry and engineering works in Diss. They made agricultural machinery and implements and about fifteen portable steam engines in the first few years of the century. Only one of these remained and I'd seen it standing in a yard near Framlingham in 1962. My offer to buy then was turned down. It was explained that there were in fact two joint owners and one of them wished to sell while the other wanted to restore. It was five years later that this was finally settled, and though I'd not forgotten about it, was surprised and very pleased to be given a belated offer to acquire.

Jack peered over every fitting like a sleuth. And when he came to the feed pump he called me over. 'Here it is,' he exclaimed gleefully. 'See, there's the punch marks I made when I did that up. They couldn't make it work and I went and fixed it where it was near Fressingfield—biked there as usual from Diss. Dot-punched the flange so I knew where it went back. I can remember the job as if it were yesterday, but it was just about forty years ago, it was. Cor, I never thought I'd see it over here, and I'm hully glad you bought it, that I am.' Then with one of his rare smiles he added, 'Besides, I don't want to run out of a job.' Jack at the time had turned seventy-six.

I too was glad I'd bought it, and local interest in its preservation led to the loan of a Young's catalogue over fifty years old, which made interesting reading. This, the last arrival of a road-wheeled engine, had not yet come in for restoration treatment, though it had been here for nine months by Whitsuntide of 1968. Its turn would come. Railways and locomotives had become the more urgent preoccupation since then. 'Thundersley' was still at Attleborough, and an unsuspected fault in the motion was taking time to put right. On one of its journeys under tow, from shed to shed, a bearing had run hot through lack of oil, and in repairing the damage, lack of precision had resulted in the refusal of a connecting rod to fit. Because the rods had never been replaced till it came to Attleborough, this now necessitated a re-bore, and a much longer task for Mr. D. W. Harvey who was in charge of the volunteer gang from the Norfolk Railway Society. It would not now be ready till late July. This delay bothered me greatly because, since I'd expected that the locomotive would be ready to come to Bressingham in early June, I'd publicized the fact. The delicate situation regarding 'Oliver Cromwell' was still plaguing my mind. Having patience appeared to me more of a penance than a virtue, and I was apt to show it at times by kicking out at those who were in no way to blame for my state of mind.

Then to make things more complicated, but adding still more to the exciting prospects—depending on the mood—came news that 'William Francis' was for sale. Apart from pictures, I'd never seen a Garratt locomotive, and indeed only one standard-gauge representative of this unique type of Beyer Peacock build remained in Britain. This was the one and it was owned by the Coal Board at Baddesley Colliery near Atherstone. As Garratts went, it was not a large one, weighing sixty tons or so as compared with the two-hundred tonners in use in Africa. But even a modest 0-4-0 ×

0-4-0 would be a wonderful addition here, I thought, if only it could be acquired.

It was Bernard Latham who told me it was for sale by tender, and I learned also that when it was retired from service two years before, it was expected to go to Canada for preservation. An agreement had been reached and the name plate had actually been sent in advance, and then for some reason the deal fell through. Now it was up for sale by tender, and knowing it to be fitted with a copper firebox, I feared greatly that the most I could offer would be far below its true value for scrap. But Captain Manisty was also aware of this, and as Chairman of the A.R.P.S. he appealed to the Coal Board for its preservation, at the same time tipping me off that Bressingham might be favourably considered if his appeal met with success. It was all very tricky, for there was no knowing which way the ball would roll. What I did know was that the scrap value would be in the region of £1,400, and that such a figure put it completely out of reach for me.

I mentioned the matter to Jim Price, a friend with an interest in steam, and a partner in a thriving local electronic components firm. He had already hinted that an investment in steam would appeal to him and in no time at all he came up with an offer of £500. This was the lowest figure I guessed the Coal Board might accept if they decided, for the sake of preservation, to forgo its full value. It would be a matter of chance whether or not anyone else, keen to preserve such a unique locomotive, would go any higher. As an afterthought I suggested another £25 might tip the balance in our favour just in case someone else plumped for a round £500 offer.

It was pointless, I thought, to go and inspect. Time was precious just then anyway. We could but take it as it was if the Coal Board decided to let it go for less than half its value. With a tender form duly completed and sent in, all we had to do once more was to wait and hope—a state of mind to which I was becoming inured with anything but easy acceptance.

Peter Manisty was the one to hear from, since negotiations for preservation would go through the A.R.P.S., and after five or six weeks' waiting his cheery voice came on the telephone once more with good news. It had been, he said, very much in the balance, but because we had good shed coverage it would come to East Anglia rather than stay in the Midlands, where some thought it should remain. And there was a hint that of the offers for preservation, ours was the highest—by £25.

Jim Price was every bit as excited as I was. We came to an agreement on joint ownership, since transport and some restoration costs would fall to me, and settled that it would remain permanently at Bressingham. At the time the good news came he was laid up, and it was Jack, Roger and I who hared off to Baddesley Colliery to size up the task of removal and subsequent restoration. No time must be lost; not only did the Coal Board want it out of the way, but I felt strongly the need to have a big loco on show, regardless of condition, since so many visitors were asking why 'Thundersley' had not arrived. The floor of the shed, now drying at last, was levelled off with ballast. On it a length of bullhead rail was laid, while Roger and a mate went off to Baddesley. It took them three days to prepare 'William Francis' for loading. The problem was not one of loading it as it stood, for this would have been easy, had I been able to afford the charge for a transporter long enough to take its forty-eight-foot wheel base. That would have spoiled the look of £1,000, but by separating the loco into three parts it would cost a fraction of this, even though it would involve three loads. It was not very difficult to disconnect the front and rear bogies. They weighed about twenty tons each, consisting each of a four-wheeled chassis topped with bunker tanks. But having disconnected the flexible steam joints, and all else that held the articulated vehicle together as a unit, there remained the boiler and cab as the centre piece with no wheels.

A frantic hunt round for standard-gauge bogie trollies was finally successful, and these Roger had taken to Baddesley. It was the task of jacking up the centre to clear both ends which took the time. The boiler then had to be lowered on to the borrowed bogies and secured for the journey. Such movements have to be planned in advance carefully, with thought given to the correct sequence, and the loading method must ensure that the object is the correct way round when unloaded. The rear portion—tank and coal bunker—arrived first and was pushed to the back of the shed. The boiler and cab, weighing a good twenty-five tons, came on a longer low loader, and wet weather was causing some worry over the approach to the shed. Low loaders resent sharp corners, and one that could not be avoided was over a culvert. Seeing the risk of the rear wheels overhanging, if not dropping into a wide ditch, we shored up with a dozen or more sleepers. The driver surveyed the route on arrival and said he didn't like the look of it, but was willing to have a go, since there was nothing more we could do to improve it. As the low loader's rear wheels cut across the corner, sleepers began to slip under the weight towards the ditch. Foreseeing disaster as it heeled over we shouted to the driver to stop, but instead he revved up and drove on. How that boiler stayed on, straining at its chains, tilting over so alarmingly, I'll never know, but it did. 'And if I'd stopped', grinned the burly driver afterwards, 'you'd have had the job of raising your engine from the bottom of that ditch. Keep going is my motto, when that kind of danger comes—but I must admit it was touch and go—I could see what it looked like in my driving mirror, and knew what would happen if I didn't make a run for it.'

23
Historic Homecoming

The wet summer of 1968 was playing havoc with the plans I'd made to be ready to receive the B.R. locos in good time. Though we were lucky on most Sundays, and gate receipts by July were forty per cent up, it was the rains during weekdays that made things difficult. The yard, sloping as it did down to the new shed, was cluttered by rails and a great heap of ballast. Rain brought rivulets of water just where space had been left for excavating. It would have been folly to lay the three tracks leading into one at the top of the slope, in the normal way, above ground level. Rail top had to be for the most part at ground level, since not only visiting crowds but also our own tractors and other traffic had to use the yard extensively. A start was made in June, but a thunderstorm came after a few hours' work, and the first trench, thirty feet wide and two feet deep, was full of water before the storm had passed.

The first cutting crossed what had been a wide ditch. We'd filled it in as useless ten years or more before with unburnable rubbish such as broken glass and old barbed wire and tins, glad of somewhere to dump them. 'Be sure your sins will find you out,' someone who had not been with us then said tritely. And out the rubbish had to come, and the hole was filled with big concrete lumps and stones after a drain had been laid to lead the water away. All three tracks crossed the same old ditch side by side, but this was only one of the causes for delay. Another had been conflicting advice regarding the levels inside the shed. This was a matter in which I was in no mind to use my own discretion, since I had no experience of big locomotives on the move. To begin with, it was suggested by a B.R. adviser that to help counteract the 1 in 44 slope up the yard, there should be at least 1 in 100 inside the shed. It needed little imagination to realize what would happen if one of the big locos got out of control down the slope. No buffers at the far end of the slope would stop it going through the wall where there was a four-foot drop on to soft ground.

It was just after the floor ballast had been carefully up-graded at 1 in 100 that Mr. Scholes came for his long-awaited inspection. Captain Manisty came too, and it was sweet music to hear their comments as they scanned the wide but empty interior, since this was just before 'William Francis' arrived. Mr. Scholes echoed the concern I too felt about the 1 in 44 yard slope, but when I told him the shed floor had now an opposite gradient of 1 in 100 he reckoned this was a mistake. 'It's not enough to check a runaway engine,' he said, 'and unless the floor is dead level you're bound to have trouble some time or other.' He went on to explain why, saying that it would never be safe to leave the wheels unchocked, and if for any reason one had to be jacked, or a man had to get underneath for some repair job, it would be dangerous. This made sense, and the next three days we spent in bringing the floor back to level.

Neither man, despite their influential position, was able to relieve the nagging anxiety I still felt about 'Oliver Cromwell'. They were emphatic enough as to their own opinion that this was a fitting place for its retirement, and were prepared to fight all the way to see that once its last run was over, it should come. But both were equally emphatic that it must still remain a secret, lest some power-wielding busy-bodies put in their spoke and forced B.R.B. to change its mind. The last run, they thought, would be early in August, which meant that although this left another seven or eight weeks for laying the turn-out of three tracks in the yard, over which off-loading had to take place, there would be no let-up for my see-sawing hopes and fears till then. When, during July, it became public knowledge that the last steam run in Britain was to be from Liverpool and Carlisle via Manchester, hauled by 70013 and a 'Black Five' on 11th August, it did not take me long to decide not to be one of the passengers. Quite apart from taking up at least two days, the persistent fear of a last-minute hitch changing the final destination of 'Oliver Cromwell' would have spoiled any thrill for me.

'Thundersley' was ready to come by late July. It was perhaps a blessing in disguise that the delay had occurred, much as I fumed at the time. But now it could wait another two or three weeks, and I could make a big saving in transport costs, by having it coincide with the arrival of 'Oliver Cromwell'. It was in the previous November that I first made inquiries about transporters able to carry such loads. The first response was frightening, with the suggestion over the telephone that the transport of even one big loco could cost at least £1,000 for the three miles from Diss station to Bressingham. Names and addresses of four other firms owning large transporters came to hand, and by May three had replied to say they had nothing of the capacity required. But the fourth had and was, moreover, keen to do the job. This was Sunter Bros. of Northallerton, and Peter Sunter who signed the letter said that apart from having the right equipment he was himself a steam enthusiast and would I let him know as soon as I could which was the big locomotive apart from 'Thundersley', the identity of which I'd been so careful in my letter not to divulge?

Some weeks elapsed before I did, because by then I was sure the secret would be safe with him. With July coming to an end, we had to press on with track-laying regardless of the rainy weather. The soil from beneath the gravel coating on the yard had to be taken away. Most of it was dumped beside the Causeway, a lane running from the low end of the farm buildings, southwards across the valley bottom. It was tree lined on both sides with flanking ditches beyond. On the east of the lane, there had been two ditches with a narrow shelter belt of trees and bushes between, but the ditch nearest the Causeway served no purpose and had over the years become partially filled with rubbishy soil from the nursery packing and propagating sheds. It was over this that I had designs for another narrow-gauge railway at some future date, and the loads of surplus soil from the yard excavation would help to raise up the formation.

A few days' drying weather at last were working for us, but George Orford on the digger tractor found the surface too hard to bite under the layer of flinty hoggin, and Buster the twelve-ton Burrell Steam Roller was brought in to help. This was fitted with a heavy Hosack Scarifier and with its spikes wound down as the engine moved it soon broke up the surface for George.

By the first week in August the top of the yard was reached. Here, eighty yards from the shed, the slope eased off where one of the main internal roadways on the place was encountered, and here the trench was over three feet deep, in keeping with the prescribed gradient of 1 in 44. There would have to be a bridge over the line at this point, but it would have to wait. This, too, was the place for the first set of points, turning out to the left for the second track into the shed for 'Thundersley', and off this the connection to the third road on which 'William Francis' already stood in the shed. The one straight line, the Number One Road, would be for 'Oliver Cromwell'.

The heavy work of track-laying brought unexpected but very welcome offers of help, as well as expert advice. Don and Arthur had by now discovered all the tricks of laying flat-bottomed narrow-gauge track, but this was very different. It was bullhead rail, in sections 45 feet long. Each single rail went 90 lb. to the lineal yard compared with the 28 lb. they mostly handled, and sections had to be lifted into position with the cranes on tractors hired from a local timber merchant. We were told how to achieve the necessary curves in the turnout by knocking out the keys, bending the rail, and then replacing the keys. But for packing up beneath the sleepers, stout shovels and crowbars were wielded by hand until one's biceps ached and compelled a pause or a change.

On 11th August, the Sunday on which the last B.R. steam trip was to be run with 'Oliver Cromwell', on its journey to Carlisle three hundred miles away, we steamed the squat little Beckton 25. Up and down it went on the eighty-yard track time after time, its twenty-two tons weight all helping to consolidate the track for the much greater weights now almost certainly soon to come. I'd been told by the Norwich B.R. men only the day before of the carefully planned timings and movements which had been so meticulously laid on for 70013 when it was uncoupled at Carlisle. Mr. Scholes had insisted that nothing must hinder its safe and speedy exit from the Kingmoor sheds once it had disappeared from the view of the thronging passengers on the train, and the sightseers at Carlisle station. They were not to know what was to happen next, what its route and destination were to be, and as it passed from section to section, it was to be guarded unceasingly every mile of the way. Mr. Scholes's chief assistant, Mr. Cogger, as well as B.R. Inspectors, were to ride footplate all the way. And once it was on Eastern Region rails our Norwich B.R. friends, who had done so much for us already, would see it safely the rest of the way to Thorpe station.

As some of the routing details came over the telephone, I began to suffer some very peculiar sensations in my stomach. My head, too, was beginning to buzz and on that Sunday I was only partly conscious of what I was doing. I drove the Beckton up and down, and for a change drove Gwynedd on several trips with passengers.

It took me a long time to produce replies to the usual questions visitors put to me, and Flora remarked when the crowds had gone that I seemed even more distant than ever. 'I haven't been able to get through to you for weeks,' she said. I said I hadn't been aware of it, even if I'd had rather a lot on my mind. 'I didn't think you knew,' she went on as we took our usual snack supper through to the lounge. 'But yesterday and today you've been miles away, and other people have noticed it too. I know what it is, but I shall be thankful when the tension passes. Have you decided yet if we're going to see Oliver come through at Harling?'

Harling Road station was about eight miles distant, the nearest point on the Thetford–Norwich line, and having been told that 'Oliver Cromwell' should be passing through Thetford at about noon next day, I'd still not decided whether or not to go and see it pass. 'I'll sleep on it,' I said, and Flora suggested that I ought to take a pill to make sure I did sleep that night. Vaguely, I was aware that I'd not been easy of late to live with, that my mind had become very one-tracked. Being confronted with the fact made no lasting difference, and as I tried harder that evening to make normal conversation, my imagination would keep on reverting to 70013, now on its way from Carlisle—unless the dreaded hitch and change of destination had occurred.

I took a pill and slept better than expected, and with breakfast over I telephoned Norwich for news. 'Due in at March in about an hour—maybe up to half an hour late at Thetford, but all's well,' I was told. 'We'll go to Harling,' I told Flora, saying that we'd leave at 11.30, and might as well make up a welcoming party of family and helpers. On the way Flora made what I thought a strange remark. 'Have a heart!' she exclaimed. 'It's the first time I've ever had to complain about you driving too fast. We've got bags of time.' But at Harling Road station the down road signal came off just as we stopped. We rushed on to the platform and I said it was a good thing I had hurried. There were a dozen or more of us, with Robert and Adrian bringing Ben, Jack, Roger, Don, and others likely to be most interested. However, the signal was not for 'Oliver Cromwell', but for a diesel-headed goods train, and when it had gone we waited twenty minutes before the bell tinkled and the signal moved again.

When the dot in the distance became a recognizable, unmistakable shape, the trouble I feared—the main reason for my hesitation in coming here—began to afflict me. I managed to hold it down, swallowing hard as this superb locomotive, quite slowly, with guards brake van behind, came nearer and nearer. I'd expected it to be faster moving than this, but by the time it came fully into view with its moving parts clearly visible, it was gliding in at only about fifteen miles an hour. I had to hold my breath, or was it that the sight took my breath away? 'Come on now,' I told myself, 'there are people about, and surely you can keep your emotions under now.' Then when the thing I so longed for was almost opposite to where Flora and I stood, the whistle blew one long musical blast, and at the same time the driver opened the throttle. That did it. I could hold out no longer. Turning away I walked slowly, head down, along the empty stretch of platform towards Thetford, and with heart thumping and throat gasping with the upsurge of emotion, I sobbed helplessly for ten minutes before I dared turn back. Only Flora would understand what had hit me, as she shepherded the others back to the cars.

It was not like that when we saw 'Oliver Cromwell' in Norwich next day. I was back to normal, or nearly so, because now there could scarcely be a hitch of the kind I'd feared so much. There might still be anxious moments, the slip 'twixt the cup and lip, but no adverse human agency I could imagine would affect the final phase. Now only twenty-four hours and twenty-five miles separated the dream from the reality. No. 70013 was tucked out of sight between the loco sheds, and though it had created quite a stir on the Monday of arrival, with television coverage and newspaper pictures and reports, it was now for a few days in the loving care of Mr. D. W. Harvey. Under his charge this and several other Britannias had come brand new from Crewe in 1951. Of the fifty-four built this one alone was likely to survive, and as we stood on the footplate Mr. Harvey told me how in 1952, by an alteration he'd worked out for the valves, a marked improvement in performance had come about, resulting in all the Britannias being given the same treatment. 'I'm delighted,' he said feelingly. 'I've looked it over and reckon it's in good shape. It makes me happy to think the only Britannia able to run has come back home, and I know it will be cared for at Bressingham. And I'd like to think that if ever you need my advice, being a steam man for over forty years, then its yours for the asking. These engines will bring a lot more people, but they'll be a big responsibility.'

This was true enough. I had become increasingly aware, for months now, how vulnerable was the state into which I had drifted in relation to estate duty and such like. It was all very well to build up the collection of engines, and it was an added pleasure for me to show them off to the public. Mr. Harvey had also said something about my being a public benefactor on this account. I couldn't accept that, but quite clearly I ought to do what I could to ensure preservation of the collection for posterity. The traction engines, except for Flora's Tidman and Robert's Sentinel, were still in my name and had increased considerably in value. The smaller locomotives now belonged to the business, but though I had dispersed a lot of ordinary shares in the Company among my family some years before, I was still the principal shareholder and life was, after all, a chancy affair.

I'd talked over my concern with both my solicitor and my accountant nine months before. What appealed most was a self-perpetuating Trust to which actual ownership of my own engines would pass, with myself as one of the Trustees for life, and my heirs to carry on when my time was up. Pride of ownership had in fact faded as the value of the engines increased, and as I began to think of them as a collection of public interest, worthy of keeping intact. Under a Trust I thought their future would be safe from the dispersal that might otherwise be imposed by estate duty. The weeks passed by, but early in 1968 at a meeting with Counsel in London my hopes of settling the thorny problem were dashed. A Charitable Trust, he said, ruled out participation both by the person making over the property or chattels, and his heirs. All must be renounced to escape estate duty, and there was no known compromise to fit such a case as this. I'd approached two M.P.s I knew who were also interested, but though they believed such a collection would be exempted from duty as being of historic or educational value, there was no precedent, as there was for paintings, old motors and a few other things. The law forced one to wait and see, and no Treasury or Inland Revenue official would give the answer, in advance of a will being proved, as to whether any part of a deceased person's estate could be exempted from duty on any grounds whatever. Nobody had so far left a collection of steam engines of public interest, and my big question was hypothetical and premature. It was very frustrating indeed, because all I wanted was to safeguard the future beyond my lifetime, and I was willing to make it impossible for the collection to be split or depleted for posterity, so long as I, and later my heirs, could retain an active interest. But since even this was not legally possible, I was advised by my solicitor to add a codicil to my will to that effect to cover the foreseeable future, and hope something more stable would be possible at a more propitious time. This had just been signed, but because the steam venture was still growing, there was also the need for some new set-up to put it on a proper footing. These B.R. locomotives—and others not wholly or legally mine—would indeed be a big responsibility. They must become, I believed, an integral part of a live steam museum collection which must be financially self-supporting.

24
Short Journeys and Long Hours

Eleven thirty p.m. was rather an unusual time for leaving home on a trip. But this was going to be a very unusual kind of trip, scarcely likely to be repeated. It was midnight when I left the car in the deserted station yard at Attleborough and the new date was 17th August 1968. I was greeted by the five men already there—made up of members of the Norfolk Railway Society and B.R. officials, who were also to make the journey. In the goods-shed, where it had been since March, stood 'Thundersley'. Flashing torches revealed a 'Thundersley' vastly different from the one we had seen at Hellifield, for now the paintwork shone brightly and all dirt and rust had been banished by the little band of volunteers who had in total given up seven hundred man-hours since March as a labour of love. Except for lagging the steam dome with asbestos cement, I had made no contribution towards restoration, and felt that although 'Thundersley' was to stay at Bressingham, still owned by B.R., these members of the Norfolk Railway Society were fully entitled to regard it as their engine.

For an hour some of us chatted while Mr. Harvey made sure that every box, bearing and moving part was topped up with oil. The night was chilly, and the waning half moon was dimly shining, as we stood waiting for the main line to clear. The mail charged through, and then came the distant throb of another diesel loco. It stopped, and after shunting one of its two brake vans on to the up line, came on to the siding and hooked on to 'Thundersley'. Then with one more shunt the little train, of one diesel, two brake vans, and a sixty-year-old steam loco between set off for Norwich at the regulation speed of 15 m.p.h. Mr. Harvey and Arnold Hoskins of the N.R.S. rode on 'Thundersley'. The rest of us were in one of the brake vans, in darkness but for a glimmer from a tiny fire in the stove.

It was for me an eerie journey. There was no need to talk in hushed tones, but we did. After all the waiting and worries and work over past months, the occasion could have been one of conviviality, but this was not the mood, which was more like that of conspirators on a deadly secret mission. From outside, above the low throb of the diesel, came the staccato hissing of 'Thundersley's' wide open cylinder cocks, since the pistons were acting as air pumps and that air had to escape. This was one reason for the restricted speed, and when we stopped at Wymondham a thorough check was made against the risk of overheating from friction. All was well, and still at 15 m.p.h. one could see in the wan moonlight misty fields of corn, flattened by the rains and storms of recent weeks which were giving harvesters no chance to begin. What had in May promised to be a year of bumper crops was quickly changing into reverse, and what was usually the driest side of Britain had this year become the wettest.

It was cold out there on the brake van platform, and I went inside to sit next to Mr. Cogger, who was carrying out his assignment to the very end. As Mr. Scholes's chief assistant his duty was to safeguard these two locomotives every inch of the way until they became my responsibility. We all chatted quietly until on rounding a bend we saw the twinkling lights of Norwich ahead. The yards outside Thorpe station were brightly lit, and here and there a man on night duty stopped to stare at our passing. When we came to a halt, the outline of 'Oliver Cromwell' loomed across the tracks and the sight gave me a tingling thrill. As we climbed down to meet others concerned, drivers and guards who were being instructed on the next stage, I could feel the mood changing to one of subdued excitement. But for me the strange air of unreality—that this was something I was seeing without in fact being there—still persisted.

With Norwich Thorpe a terminal station, a double check was made to ensure correctness of the order in which the train was made up. It had to tie in not only with loading preparations already made at Diss station sidings, but with those for off-loading at Bressingham too, so that they would stand tender first to the new shed. At last all was ready, and as the little train (now longer by 'Oliver Cromwell's' bulky seventy feet) set off again, I was on the footplate. Strictly it was against the rules, and it was not until Mr. Cogger was about to climb up that I asked if I might join him. He was willing enough, though his authority may not have covered such a request. But since no one wished to deny me such a pleasure, up I went with the promise to make myself scarce if the need arose. It was over seven years since that memorable footplate ride to London on 70034 'Thomas Hardy', and to ride now on 70013 'Oliver Cromwell', actually on the way to Bressingham, was about the only way of proving that what had seemed so incredible was indeed coming true.

With a double set of cylinder cocks hissing in loud rhythm we gained the rise on to the main London line. Over the hill of Mousehold Heath beyond the still sleeping city, with its slim cathedral spire stabbing the dim sky, dawn was breaking mistily. As we rounded the bend, into a cutting and under the main road bridge, I wondered if the lights of a car coming towards Norwich would pick out this strange cavalcade as we came out on to the embankment. Had I been at the wheel I would have jammed hard on the brake to watch and wonder, yet how much more exciting it was to be where I was. A herd of young cattle farther on saw and shied hastily away with tails in the air. Pigeons flapped noisily from coppices which echoed the sharply hissing cylinder cocks. The escaping air was under such pressure that we could see it, almost like steam, as the day grew lighter. The dawn air was sharp and cold on my face as I leaned out to watch the fascinating motion of the Walschaerts valve gear until I had to draw back.

We were running late, but stopped for a check at Tivetshall at 5.30, the time arranged for arrival at Diss. Some unexpected delay at Norwich had caused this, but there was to be no exceeding the speed limit Mr. Harvey had imposed, and in any case we should still clear the main line at Diss before the daily timetable began. The line was dead straight now and by six o'clock it was light enough to see the outlines of Diss station, and the big new granary nearby, from a mile away. Little groups of people had edged up the line to meet us and cameras were clicking long before we drew to a halt, even though the light was poor. When I stepped down from the footplate, it was as if all the anxieties of past months had never been. I'd missed a night's sleep, and was thankful to have done so, though I was becoming very hungry. So was Mr. Cogger, and after watching the shunting manœuvres into the prepared siding, we both went back to Bressingham for a hot breakfast and a short kip.

One of the arrangements made beforehand was that of guard duty. With 'Oliver Cromwell' and 'Thundersley' now my responsibility, having arrived at Diss, it was my duty to keep to the maxim applied by Mr. Scholes and the B.R. officials, that nothing must be left to chance. With Robert a Round Tabler, a roster of volunteer members had offered to stand constant guard with the recently retired P.C. Edwards as night-time reinforcement. The prevalence of thefts by souvenir hunters was not going to spoil these two, and when I returned to the station later that Friday morning a pair of volunteers were already on the job, keeping an eye on the sightseers, even though these were mostly local folk who had come solely because they were interested.

Sunter's big transporter had been delayed on its way from South Wales, but Peter Sunter himself arrived from Northallerton to supervise. The route had been surveyed by their foreman some time before, but only one doubt remained. The transporter might be a few hours late, but the real worry was whether with 70013 loaded, it would clear the bridge carrying the railway over the main road leading to Diss town and beyond. Bressingham was only three miles away, but the headroom under the bridge was only sixteen feet. It could jam drastically and if it did not clear it would mean a twelve-mile detour to Bressingham. In spite of rainy weather persisting, this road took considerable holiday traffic, and Friday was market day in Diss. The police had of course been consulted and were prepared to co-operate and provide escort. But their request to confine movements to the less busy daylight hours were not to be ignored, and the delay to the transporter set up an unexpected worry.

'Thundersley', the lower of the two, was to be loaded first, and while we waited, Peter Sunter and I measured the headroom once more, borrowing a long spline of wood from a nearby merchant's yard. This had been done before, but with 70013 far away, dead accuracy on its height was difficult to attain. We found the distance from the road surface to the bridge varied by up to three and a half inches, and even allowing for all adjustments possible on the transporter platform it would be touch and go when it came to the vital moment. Peter believed it would just clear, and with the transporter at last in sight, there the matter had to rest. If 'Thundersley' was to be loaded ready for movement by dawn next day, loading had to begin at once. It began to rain again as the huge thing with its sixty-four wheels, hitched to the powerful tractor, backed over the siding. Another of Sunter's lorries had arrived. It carried great baulks of timber and these were unloaded to build a ramp forty-five feet long between the waiting 'Thundersley' and the rear of the trailer.

The hours passed and the rain came down harder than ever. The ramp was ready by 8 p.m. but it was too late to begin the slow, tricky pull from the tractor winch which was to draw 'Thundersley' at snail's pace up the ramp and on to the transporter. The day which had begun so full of promise had ended dismally, and the timing programme so carefully prepared had gone by the board. I'd tried so hard to see that police requests were complied with, and there was the danger now that instead of having both locos safely at Bressingham by early Sunday morning, we might not avoid running into trouble with the afternoon crowd of visitors.

With a new day begun, we were back at the station at 5.30 a.m. 'Thundersley' began its slow rise. Fresh fears arose when most of its weight fell on the front bogey mounting and the driving wheels became suspended above the rail. It was not until the rear wheels allowed the angle to close, as these too touched the ramp, that the danger of damage was past. But once clear of the siding rails, the backward tilt of the transporter platform helped to spread the weight once more. Peter Sunter had been confident all along and he had been right. This was the right type of vehicle for the job. With the tractor, the outfit had cost his firm £45,000, but it could carry loads up to a hundred and fifty tons and lengths up to eighty feet.

'Thundersley' had taken five hours of the morning by the time it was fastened down and ready to move. Down the slope, lined with spectators, we went, and then sharp right to go under the bridge. Once there we paused and edged slowly forward as the transporter's donkey engine lowered the platform so that the engine's steam dome, its highest point, cleared. Measurements were taken that might help when it came to 'Oliver Cromwell's' turn, when this same bridge might also become a bridge of sighs. Yet it was all so easy to turn right-angle bends with those sixty-four wheels in rows, all capable of being steered themselves. Along the straight length of Victoria Road more people stood, with lines of cars held up. From the footplate I could look down upon them, and I could smile because now I felt like smiling. Someone I had expected but did not see was a B.B.C. cameraman, having previously given them the tip. Anglia Television could not cover the event because of the technicians' strike, and later I learned that the

B.B.C. missed it because their cameraman could find no film.

Here and there the driver stopped to allow cars to pass. But just behind, like a cortege, came a line of cars including ours with Flora at the wheel, that were following all the way. Children on bikes or on foot rode and ran beside and behind as we edged slowly along, and by noon we were outside the nursery entrance just as another heavy shower with thunder passed overhead. The entrance from the road had just been specially widened, but the transporter's wheels sank in deeply. They churned out some sleepers we'd laid to widen the next right-angled bend and one of them burst an air pipe which, by leaking, put on the brake. There was another delay from the same cause before finally it came to rest over the newly laid No. 1 track in front of the shed. With the timber merchant's crane tractor to help, the ramp-carrying lorry was unloaded once more and by teatime 'Thundersley' was safely off. Arthur had been waiting with 'Bertha' in steam, and as the ramp was re-loaded he drew 'Thundersley' up on his winch, to clear the points. Then with wire rope still taut, and with the locomotive hand brake under control, it was eased gently down the slope into the shed on No. 2, the middle of the three tracks, to stand next to 'William Francis' on No. 3.

When we got back to Diss station after tea, half the local population seemed to be there to witness the final act. A rope had been strung across to keep the loading space clear, and two policemen were on duty as well. Peter Sunter's six assistants worked hard to build up the ramp so that 'Oliver Cromwell' could be raised before dark. But as dusk fell, there was still much to be done, and a night's rest was obviously needed. The police understood, and seeing how we were placed relaxed their first injunctions, saying that as 'Thundersley' had caused no serious traffic problems on the Saturday, an early start next morning was not so vital.

What still remained vital was that 70013 should clear the bridge, and when at 11 a.m. next morning 'Oliver' stood gauntly and bereft of his tender on the transporter, it seemed impossible. Onlookers must have thought so too, for they swarmed up the embankment with cameras ready as we made the approach. With me on the footplate were Robert and Adrian and at the crucial moment all three of us, as well as Peter Sunter, were out on top where at eye level we could see what was happening, and shout if need be. The platform fell to its lowest possible height. The squat chimney of 70013 cleared by an inch or so and inch by inch we edged forward. At one point a piece of bridge guttering had to be forced up clear of the dome, and then at last the right away signal was given to the driver. The crowds drew back and we stopped to let the traffic jam disperse. 'Phew!' exclaimed Peter. 'If they'd given it another coat of paint we shouldn't have made it.'

But we had made it, and the crisis was over. With elation mounting, I wished some means of blowing the whistle could have been found, or that the procession could have been headed by a marching band. Since the episode at Harling last Monday I'd not been assailed by such emotion again, but now the event which I had dreamed, schemed and fretted over for so long was reaching its fulfilment. It was not pride of achievement, nor of ownership, but sheer happiness from the good fortune that had brought me into this position, to be the chosen guardian of this noble locomotive. It made me proud to have my two sons up there with me, who had shared the lean times past. They did not also share to the full my enthusiasm for steam, but we were close at heart in our love for all Bressingham stood for, separately and together.

As we came through the wooded S-bend where the road from Roydon slopes downward to Bressingham, and where our land began, my mind played a trick on me. It flashed back to March 1950, when on a borrowed bike I coasted down that slope for the first time after my return from Canada. Then I was nearly broke, with nothing but hope that I could find the will to begin again, to restore what I'd so nearly lost, knowing that only years of hard work would suffice. That was the day I came across the few useless remains of 'Bella', my first steam engine. I was alone on that first sad, chastening visit because, with the house tenanted, we had no home to re-enter together. But now, wishing that my whole family were up there with me, I could from this height see over field after field of flowers and crops, and that elevation was gained from the cab of the one locomotive, above all, I could have wished for if such an unlikely bait had been offered. Truly, life had treated me so much more kindly than I could ever have hoped, or for that matter deserved. And with two fine sons beside me to carry on, it was small wonder that I felt proud and happy as I waved and smiled down at Flora with Anthea and Jenny beside her, in the car following close behind. And Bridget was with them, having come home from London specially that weekend for the great event.

25

Cold Water and Hot

If great care had to be taken when unloading 'Thundersley', even more stringent precautions had to apply to 'Oliver Cromwell'. For one thing, it was nearly forty tons heavier, even without the tender (on which the only hand brake was fitted), and for another its roller-bearing axle boxes made for freer running which could well be a disadvantage in present circumstances. The third factor was that it was Sunday afternoon, and even if no more than an average crowd turned up, most of three thousand people would want to witness the operation.

The ramps were in position just at the opening time of 1.30, and very soon helpers were being mustered to keep back people who had little notion of danger. A wire rope could snap and spring back to cause injury, apart from any other result of such a breakage. The operation was based on such tension to offset gravity, and since 70013 had to come off the rear of the transporter, the gradient was against the vehicle's winch. This had pulled it on, but one of our steam engines had to pull it off. Arthur was well up on the parking meadow with 'Bertha's' hind wheels blocked against the strain as her winch wound slowly in. For safety, the transporter's winch cable ran out as Arthur pulled in, but the co-ordination needed between the two was difficult because of different speeds.

I had 'Black Prince' also in steam, knowing that a double cable would be needed when it came to letting the locomotive down the slope to the shed. Standing close to the track against the highest set of points, I saw how 'Bertha's' cable twanged and strained as 'Oliver Cromwell's' weight touched our rails. With its six-foot driving wheels suspended, bridging the angle of the ramp where it met the track, it looked as if something must snap. Something did, but it was 'Bertha's' cable, which snaked back with its jagged, broken end lethal for anyone it hit. This took place only a few feet from where I stood with 'Black Prince', and spotting the reason I asked Peter Sunter to put his winch into a higher gear, so as to match more nearly with 'Bertha's' speed. Arthur had pulled as slowly as he could, but it was too fast for the releasing cable at the rear and 'Bertha' had snapped the rope.

Ten minutes after pulling began again, 'Oliver Cromwell' was down, with all twenty-four wheels on Bressingham rails and safely chocked. Now that I could relax, I climbed up to its cab and smoked my pipe as if I'd not a care in the world, while the ramp was dismantled for the last time. Sitting in the driver's seat I was almost out of sight, and felt no qualms for what remained to be done. Almost oblivious to the throng below that filled the square of the yard from Flora's gaily whirling Gallopers to new shed, the Dutch barn, and back to the narrow-gauge railway station, I was wondering how quickly we could complete the excavations and lay the remaining one hundred and fifty yards of track through the parking meadow.

The transporter moved from its straddling position over the line. Beyond, just in front of the open shed door, stood 'Oliver Cromwell's' tender. This had been brought from Diss station earlier that morning by an independent low loader. Now it was 'Black Prince's' turn to work and, changing footplates, I stoked up, blocked the rear wheels, and set the throttle as the cable was laid out. Of the road engines 'Black Prince' was always my favourite and no one else drove it unless I asked them to do so. Its gears were noisy, but it still had the power, in spite of being older than I am. Down went the rope, pulled by helpers who shouted, 'Keep back please!' Robert stood on the waiting tender, on which was the only manual brake a Britannia possessed. With the rope hooked on, I threw the lever forward and 'Black Prince' began to pull. It was easy work. All that mattered was to keep people back from the rope and ensure a steady pull.

'Bertha's' cable was still hooked to 'Oliver Cromwell's' for safety's sake, and when complete once more with tender attached by its safety chains, chocks were drawn out and '70013' was allowed to run back down the gradient, but only foot by foot. This bothersome gradient of 1 in 44 was a kind of bogey by now. We still had to prove its effect under varying weights and circumstances, and total caution was called for, since there were doubts whether the handbrake alone could check the hundred and thirty tons' weight. But with the cables of 'Bertha' hooked to the front and 'Black Prince' fixed to the rear, with four helpers sliding chocks each side on the rails just ahead of a wheel, as well as Robert's hand ready up there in the cab for the brake, there was no real danger.

It took half an hour for 'Oliver Cromwell' to gain the level, just inside the shed, where the gradient ended. For the rest of the way —the engine's own length—four of us with pinchbars levered the wheels and the move was complete. I went across to Flora on the Gallopers after that. We'd not met or spoken together since the hasty lunch four hours ago. 'That's that,' I said cheerily, 'no snags, no trouble and all's well.' She gave me a wan, worried smile. 'I've been having kittens,' she confessed. 'All I could see from here were crowds of people getting in the way. It all looked so terribly dangerous, and I felt so scared that somebody would get hurt.' 'But everything was under control,' I protested, rather taken aback. I explained all the precautions we'd taken, how nothing had been left to chance, but looking across I too could see how from this angle it might well appear as if people had stood dangerously in the way. 'If only I'd known you were using two engines and ropes,' she continued. I'd have made time to tell her, of course I would, had it occurred to me that tied as she was to her job on the Gallopers, Flora might worry. 'Just like a man,' she added finally as she realized and forgave yet another of my omissions.

The crowd that Sunday had not been larger than usual, because so few people knew that these two splendid additions were coming. With no television coverage to spread it about, I doubted if for the rest of the season their advent would swell the gate receipts. These were well up anyway on last year's record, and without depleting last year's total donations to the usual charitable causes, there would be enough to make a considerable one to the Transport Trust. As a registered charity the Trust was acting in line with its objectives by paying the cost of transporting these locomotives from Diss Station—costs which were well beyond my means. The Trust had also been appointed as custodian on behalf of British Railways Board to safeguard their future on loan, to Bressingham Steam Museum. I had consulted some of the voluntary helpers, two of whom acted with me as Trustees of the Charity Fund and whose signatures were needed for disbursements by the Fund. They were as convinced as I that these locomotives would be an investment and bring increases in gate receipts far in excess of the outlay.

After the tension and excitement, the annual 'Steam Gala' was not an event I was looking forward to. Though Mary Fox, the firm's Secretary, relieved me greatly of the detailed arrangements, the final responsibility was mine. Robert could promise no help, and because harvest was so long delayed with crops flattened and still too wet to combine, the event was to him nothing but a nuisance. Until last spring I'd been half inclined to call it off, but had relented because so many people seemed to enjoy it. Coming as it did at the end of August, it could scarcely avoid conflicting with harvest now that combines had taken the place of binders with sheaves, stooks, stacks, thatch and threshing drum. I sometimes wished combines had never been invented, for they had destroyed the sweaty glamour along with the labour of harvest. Perhaps I was old-fashioned, but Robert was certainly a modern and at the mere suggestion from me that we might harvest one field in the old way I was met with an amused but scornful rejoinder, 'So that you can thresh out with the drum and a steam engine I suppose? Well, if you can provide the labour and keep the extra cost from going on to my farm analysis record, go ahead—you're still the boss,' he would say. As it happened the 'Steam Gala' this year was over before more than ten acres of wheat were combined, and still it rained sometimes every other day, and often every day, little or much.

Rain did not spoil the event which took place a fortnight after the arrival of 'Thundersley' and 'Oliver Cromwell', but there was very little sunshine either. The crowd of about eight thousand each day was quite enough to cope with and, though I felt inclined to put it over the microphone that this Gala would be the last, I refrained, knowing that I might again be persuaded to keep it up as an annual event. The hectic summer and added responsibilities were showing me a glimpse of the red light as warning that I could not expect to be able to keep up this pace for long. The steam side, along with public openings, had grown so much that now I should have to pause and take stock as far as management and organization were concerned. It was becoming almost a business in itself, yet in many ways was intertwined with farm, nursery and garden. Something would have to be done before long, and the need for a manager seemed to be indicated.

It did not rain on Saturday, 14th September, but it was dull, windy and cold. I had looked forward to this date, because it was booked for a meeting here of the Association of Railway Preservation Societies. About ninety delegates turned up and apart from giving rides on the roundabouts and the Fen Railway, we laid on refreshments. It was a pleasure to meet and talk with others interested in preservation, and for the first time I was given a full insight into the very useful work done by this Association, under the Chairmanship of Captain Peter Manisty. The meeting itself took place inside the new shed, with 'Oliver Cromwell' looming hugely nearby. The shed was beginning to fill up with engines by now, but the three big locos had not taken up all the centre space. There was in fact a fifty-foot length on the same line on which 'Thundersley' stood, and the suggestion had been made that No. 2500 would make a worthy addition to fill that space. This was the class built by Sir William Stanier in the 1930's to replace the 'Thundersley' type on the L.T. & S. Railway. All thirty-seven of them were 2-6-4 Tanks with three cylinders and Walschaert's valve gear, and to my mind were the most handsome of all tank locomotives. No. 2500, being the first of the class, had been scheduled for preservation, and retired from service in 1962. Now it was at Brighton, and so far as anyone knew there it would stay until some museum offered to take it over for restoration and exhibition.

Captain Manisty was enthusiastic when I put the suggestion to him that Bressingham would be a fitting home for the '2500', and he promised to back my application. 'Why not try for the J.17 too?' he added. That was a Great Eastern engine and had been restored already by B.R., and nobody had put in for it yet, so far as he knew. If I did not jump at the idea it was because I wasn't sure if there would be room. It still remained to be seen if the fourteen traction engines would fit into the space allowed for them, and most of them were still outdoors. Another thing was that this 0-6-0 goods engine with tender was inside-cylindered and less attractive to me on that account. And it was the very same type which figured in the childhood nightmares I used to have of being fixed to the line as it approached with its extended piston rods moving in and out so menacingly as it came closer and closer.

Next day it was still dull, cold and windy and we expected no large crowd. At lunch time it began to rain with the wind just south of east, and it fell down pitilessly all afternoon. 'George Sholto' went round the track once and was put away. Gwynedd stayed in steam at the station, but no passengers seemed to be interested. Such few people as had turned up were nearly all in the big shed, and Flora played the organ to keep some air of jollity during what was turning into a very depressing afternoon as the rain pelted down. When Percy took the readings next morning it was still raining, and from the container came out a record amount of

4·38 inches—double the heaviest twenty-four-hour fall I'd ever recorded over a period of nearly forty years.

This spelled trouble. Through our narrow shallow valley had to pass all the surplus water from an area of about thirty thousand acres. On every acre over four hundred tons of rain had fallen, and with the land already saturated it was bound to bring a flood. That morning we found a culvert blocked by debris brought down the stream leading to the big pond. Several frames recently filled with over a hundred thousand cuttings were submerged. Water had eroded the slopes, washing plants bare of soil in some places and covering them with silt in others. Patches of water were already standing here and there in most of the fen fields, and by nightfall nearly a hundred acres were covered. In two of the fields, half a million plants in full growth were slowly being drowned. These fields I'd skirted when driving Gwynedd round, with passengers behind admiring the colours, had been a delight to me, for as a nurseryman my appraising eyes were studying quality rather than colour. The autumn lifting season had begun two days before. Orders were flush and quality had never been better. But now, unless the flood receded quickly, many of these plants would die and many of those orders would not be fulfilled.

On the Tuesday, it rose higher than ever. I tried to walk round on the elevation of the Fen Railway. In places even the rails themselves were under water and here and there a washout had occurred leaving no ballast beneath the sleepers. Of course we did what we could, but there was so little that could be done. The river could not take it, and until its level subsided, we could not begin to induce the water to leave the land. Such a flood had been a fearsome hazard at the back of my mind for years, in the belief that whatever weather adversity had come in the past might some time or other recur. Though rare, such deluges had fallen before in East Anglia, but the last one comparable with this was in 1912. I'd been pleading for a year or two with the Drainage Authority to dredge the river, and though in response a survey was being made it held little promise, and was of no help now that a flood had come. And it had come at about the most damaging time, since plants are much more easily killed when in full growth than when dormant.

It was ten days before the black soil surface could once more be seen. By then, what had been lush green growth was sickly yellow, rotting and stinking. For days, as soon as the water began to recede, several of us had been ranging the fields with spades, making little channels to the nearest ditch to hasten its passage. But in the office they were typing letters to firms who had placed orders with us, explaining why this, that and the other could not after all be supplied. This was an ironic aftermath. One of the reasons why we were the largest wholesale producers of hardy plants was that we had such a wide variety which none of our competitors offered. Now that we had lost several kinds we could not buy in to make orders on hand complete, but it was poor consolation to know that our customers could not do so either.

The Fen Railway was out of service the following Sunday because of the flood. But, by way of relief, I decided to steam 'Oliver Cromwell'. The rest of the track had been laid just in time, the most exciting steam-up ever was planned for Sunday, 22nd September. To avoid fouling the new shed with coal smoke, any engine had first to be drawn out before its fire was lit. Six hours, or maybe seven, were needed, or so I was warned, to raise a head of steam from cold. That would mean, I thought ruefully, rising at 4 or 5 a.m. and still the loco would have to be hauled out first. A winch had been thought of, but for this first occasion I decided to steam 'Black Prince' overnight, and with fire stirred up by about 5 a.m., 70013 could be drawn out by steam, and with luck it would be ready for a trial run before opening time.

In pitch darkness that morning I raked over 'Black Prince's' smouldering embers, and by 5.30 a volunteer who was camping overnight came out to help. The shed doors were slid back, and slowly 'Oliver Cromwell' emerged at the end of the cable until the cab was clear. I'd prepared several barrow-loads of wood in advance, and with the kindling in the 6′ × 7′ firebox, diesel-soaked rags set it off. Then came billets I'd split in readiness from an old fallen willow tree, and these turned to flame as fast as we could throw them in for the next fifteen minutes. It was smoky but warm in there, and when coal too was shovelled in, 'Oliver Cromwell's' chimney was belching smoke at a rate which proved that the tubes had been well swept. Between times the Beckton 25 was also lit, since I thought it might be wise to run the two together for extra braking power should the gradient prove troublesome. This unproven factor was about to be tested, and no risks must be taken.

At 8.30 the pressure gauge stood at 50 lb. and soon after that I tried the whistle, and the sweet low sound was music indeed.

Hurrying back to the house for breakfast, I found Flora and the girls having theirs. 'We heard it!' they exclaimed excitedly. After that, having somehow nearly halved the usual B.R. time for raising steam, it was a case of keeping the fire low till George Ewles arrived. As a recently retired Norwich B.R. driver, and a lover of steam, he had gladly agreed to come in response to the B.R. rule that only approved drivers should handle their locos.

He stood admiringly back for a moment, smiling broadly and saying it was good to see an old friend again—one which he had often driven at ninety, and once touched a hundred miles an hour he reckoned, on the Norwich–London expresses. 'And we could always get top speed through Diss. Well, let's have a go, shall we?' It had been laid down that to begin with no one but driver and fireman or B.R. official should be on the footplate, at least for that day. I watched from below as a shower fell. With cylinder cocks open George pulled the regulator, and the great engine moved easily up the slope. Beyond was a curve across the meadow, and here it lost grip and nearly stopped, though the wheels turned faster. It gave me a nasty feeling in the pit of my stomach, but George Ewles leaned out and grinned. 'Wet rails—she'll be all right after a bit.'

She was, and though Flora too had become depressed at the thought of my not being allowed on the footplate, I couldn't believe the ban would last for long. By soon after opening time it was relaxed, sure enough, and up I went. I waved gaily to Flora, over there on her Gallopers, as we set off again on the short run, blowing the whistle for sheer joy. After two more runs I was called away to sort out some parking problems, and when I came back, 70013 was on its way back to the shed again. And there on the footplate stood Flora herself, radiant with excitement. She too had been invited up while I was absent. Climbing down she stood there almost gasping as she tried to tell me what a marvellous thrill it had been. 'I'm lost for words,' she said. I could well believe it, having felt that way myself.

26

A Museum is Born

As a teenager, Verse 10, Chapter 9 of Ecclesiastes took my fancy, and to a large extent has been a rule of life for me. 'Whatsoever thy hand finds to do, do it with thy might' seemed pretty sound advice, but after fifty years striving to do just that, I could look back and see some of the consequences. There have been occasions when I became over-absorbed, paying so much attention to projects and things that I tended to neglect people. By 1969 one or two danger signals could no longer be ignored and it was obviously time to slow down and take stock. The concept of a live steam museum had taken shape, especially during the previous years, but it had taken a toll on family relationships, and to some extent on both the nursery business and the garden I'd made, without which the steam museum would never have developed. As it dawned on me just how deeply I was committed, the need for someone to relieve me became more and more insistent. I was not an engineer and it was obvious that the only solution lay in enlisting a qualified man to take charge of the restoration and maintenance work which was piling up. Geoff Sands was such a man, and he agreed to come in the autumn for a much lower salary than he was being paid by B.R., just to be with steam again.

That summer opening season proved a vital point. The number of visitors shot up, as it became known that 'Oliver Cromwell' was in service every Sunday, giving footplate rides. This in turn increased narrow gauge and other takings so that, from the financial angle, we could be self-supporting, with a regular and competent staff.

Just in time for a 'Flying Scotsman' excursion to Diss and Bressingham, the new 'Woodland Line' was completed and in operation. Though this exciting day was marred by drizzly rain, it was during a thunderstorm that 4472 and its train stopped at Cambridge that evening on its way back to London. By invitation from Alan Pegler, I'd boarded it at Stowmarket and rode footplate most of the way to Cambridge, to be met by Flora with the car. Though this in itself was stimulating, it was somewhat in conflict with my efforts to avoid being over absorbed. But I was already up to the neck, and if some help in consolidation was vital, I could do no other than go on. It came as something of a surprise to realize that Bressingham was now the most comprehensive live steam museum in the country, and yet it was equally plain that much more had to be done to make full use of the scope which existed. The setting and the potential were here, waiting to be exploited.

The number of visitors during 1969 increased by forty per cent, and many of them queued to be given the short footplate ride every Sunday afternoon on 'Oliver Cromwell'. Many more were fascinated to see it in motion, and there was no doubt at all that this was the main attraction. It was still my job to light up, but I soon found that steam enough to use the blower could be raised in less than three hours. In less than another hour, by which time the rostered B.R. driver had arrived, it could move under its own steam. And with cylinder cocks open on the one in forty-four gradient this first dead slow movement was a thrilling yet awesome sight.

This came as some compensation for the task of lighting up. The procedure was to light oily rags on the firing shovel, and follow up with dry split kindling, gradually working in larger pieces of firewood until the three or four barrow loads stacked in readiness overnight had disappeared into flames. Then I began shovelling in coal. This very quickly produced smoke, with as much or more coming back through the firehole door into the cab, making me gasp for clean air, as it found its way through the tubes into the chimney. But having found out that it took about sixty shovels full of coal to cover the flaming wood there was no sense in closing the firehole door until this amount was in. I could then go indoors for a leisurely breakfast. No more coal would be needed for at least an hour.

After breakfast, with other Sunday helpers arriving, smoke would be coming from other chimneys too. One of the little Becktons, with David Ward in attendance, was in service as contrast to 'Oliver Cromwell'. Three narrow gauge engines at the rear of the museum shed were also being prepared whilst in the open yard, two traction engines in turn were put in steam each

Sunday, along with the little 'Princess' on the 9½in. 'Garden Railway'. It suited me to drive one of the 2ft. gauge, and now the 'Nursery Line' had been extended once more, it was a pleasant run. My choice was for 'Bronwllyd', the 0–6–0 Hudswell Clark newly restored and in service, and no matter how many train-loads of visitors I hauled it was never monotonous. There was craftsmanship in keeping up steam, in a judicious use of the regulator and in treating one's engine as if it were human. It was a craft in which proficiency was reward enough, and even if I was still an amateur, I could understand the intense pride that came to the professional driver as he came up the long, hard way to become a master of his art. This was an art which only steam power could foster.

During that summer, on Thursday and Sunday afternoons, I drove 'Bronwllyd' on the 'Nursery Line', now extended to just over one and a half miles, over five hundred times. An exchange of land with a neighbour took place during that time, and this made possible a further extension to make it just over two miles. Winter came and more track-laying began. But though the terrain was reasonably level, and mostly on sandy soil, one stretch was over the heaviest alluvial clay imaginable. A large new dyke had been dug, and the huge lumps the excavator had thrown out had to be broken down and levelled off before sleepers and rails could be laid. It took a tractor and trailer half-loaded with sleepers an hour to travel 150 yards because of the mud and clay, and in such conditions much hand labour with planks and barrows was needed to finish the job.

Meanwhile Geoff Sands had arrived to take charge of work inside the big museum shed. The ex L.M.S. No. 2500, 2–6–4 Tank came in for stripping, scraping and repainting, and under Geoff's expertise as a qualified Loco Engineer, 'Wm. Francis', the much neglected Garratt, was brought back to respectability. This unique locomotive had been something of a headache. With its three detachable sections joined together again, I had decided the year before to raise steam as a means of finding out its internal and mechanical condition. But though with much difficulty some steam was raised, it refused to budge any of the moving parts. Instead, it escaped from various places which should have been steam tight, and with the fire drawn the engine was placed where a complete overhaul could be begun. It was weeks before the grime and rust gave way to primer paint, involving Jim Price and his helpers as well as ourselves. The steam test had proved the tubes to be leaky and useless, and when all these were knocked out the purging of boiler and firebox yielded one and a half cubic yards of shale—enough to prevent not only effective steaming but to risk violent damage through lack of firebox water space.

The cost of restoring 'Wm. Francis' was more than double its purchase price, by the time new tubes were fitted, its mechanical defects put right, and its body work fully preserved with several coats of paint. I was beginning to learn that the gulf between reality and challenging visions of an engine restored from a state of dereliction was apt to be much wider and vastly more expensive than allowed for in my early optimistic enthusiasm. But if I failed to make sufficient allowance at the outset for the unexpected snag or defect, the joy of seeing such an engine come to life again, fully restored, was all the greater.

The beginning of the 1970 opening season soon revealed the need for a second train run on the 'Nursery Line'. Though the circuit took no more than about ten minutes, at peak times visitors were queuing for tickets, and it was bad business to keep them waiting. 'George Sholto' coped with the smaller numbers using the less conspicuous, shorter 'Woodland Line', and with more toast rack coaches hurriedly built 'Gwynedd' was brought in for the second train on the 'Nursery Line' on Sundays when most visitors came. This meant that we had no 2ft. gauge locomotive in reserve and having heard of one for sale in Germany I made a hurried trip to Bremen, with David Weston for company. As an artist specialzing in steam he was up and coming, and having done a deal for a rather ugly, heavy Orenstein and Koppel, we visited the Hamburg loco. sheds, as well as a collection of disused steamers in Denmark.

As a trip it was stimulating, but its outcome was not effective in meeting our needs for a reserve locomotive; nor was it conducive to my keeping my priorities right. In the latter sense, the steam bug became over-active once more, and when the O. & K. arrived its defects were numerous enough to show up my lack of perception and knowledge. It took weeks of repair work, apart from restoration, before it became a runner, and when at last it was brought into service late that summer it proved also to be too heavy and too long in the wheelbase to negotiate some of the curves of the track. Though powerful enough to force its way round them, the effect was to widen them so that coach wheels were apt to lose contact with the rails and drop in. This was not dangerous, but it was bad business even if most male passengers were willing enough to heave a vacated coach back on to the line as if it were part of the fun of riding behind steam.

Apart from these relatively minor troubles—overcome by keeping the O. & K. strictly in reserve—1970 gave further proof of public interest and support. Gate receipts were well up again, and with no major restoration tasks on hand I was seriously considering acquiring one of the redundant Danish locomotives I'd seen. The Germans, I'd discovered, were against parting with any of theirs. Their policy of phasing out steam gradually meant that only life-expired locomotives could be considered for sale, and one had to admit that this made much more sense than the haste of B.R. in the switch to diesel involving the disposal of engines capable of further running for years.

The Danes had a less vital rail system, and some attractive Pacifics as well as 2–10–0 German-designed freight engines, just taken out of service, were for disposal, all at reasonable prices. But for the frightening cost of shipment, I would have taken the plunge during that autumn. Geoff Sands was willing to go over and pick out a good one, and a date was fixed for him to do so when two deterring factors emerged. One was that it would cost about £3,000 to load and ship, and the other was a call Geoff made in October to see 'Royal Scot' at Butlins' Skegness Holiday Camp. 'I'd dearly love to see it here,' he told me. 'The man in charge says it's a nuisance—standing there rusting away with people tampering with it—an eyesore rather than an attraction.'

An attraction it no doubt had been when Sir Billy Butlin had the idea of putting some ex-B.R. locomotives as static exhibits on four of his holiday camps. But that was in 1962, and after eight years serious deterioration had set in. I'd been told what was happening long before this, and a letter I had written to the owners in 1966 suggesting taking one over for preservation had met with a rather blank refusal. With considerable diffidence I wrote again, and the reply was such that I guessed the Butlins' Board of Directors must themselves have come to the conclusion that it was time their locomotives moved inland for preservation. The upshot was that in November 1970 one of the Directors visited Bressingham, and having looked us over said we could take our pick, and so long as we paid all transport and restoration costs and maintained the locomotives, they could come to Bressingham and stay there permanently.

This posed a problem. Two facts stood out sharply. One was that four large locomotives and four smaller ones all in need of extensive, costly restoration would be more than we could cope with. The other was the near certainty that other preservation centres would dearly like to share them. I felt, too, that it imposed a burden of trust too great for my shoulders alone to bear, and if we had our pick then the Transport Trust should have the responsibility of placing the rest elsewhere. But having set things in motion after choosing the locomotives at Skegness and Ayr Holiday Camps for Bressingham, the news leaked and some preservationists reacted with wolfish ferocity. It also turned out that because the locomotives sited at Pwllheli and Minehead Camps were featured in Butlins' 1971 Brochure, which was already printed, these would have to remain where they were for another year to avoid contravening the Trades Description Act.

For several weeks it was touch and go whether or not we would be allowed to take those at Skegness and Ayr. Others tried to make out cases of prior claim which were no small embarrassment to the owners, the Transport Trust and to myself, along with my helpers and well-wishers. At last, towards the end of February, the way became clear. The most vociferous of the anti-Bressingham enthusiasts had managed to blot his copy book, and Peter Sunter was ready to get the wheels of his big transporter turning. 'Royal Scot' was hauled by road from Skegness. As the crow flies, it was well under a hundred miles, but the need to avoid weak or low bridges entailed a distance of over two hundred, and took three days en route. The 40-ton L.S.W. Shunter, also at Skegness, was much less difficult, but by the time both had arrived so had trouble with regard to 'Duchess of Sutherland' from Ayr.

This 164-ton Duchess Class had been expertly prepared for hauling by rail by Mr. Bill Bennett of the Ayr M.P.D. of British Rail. With its four cylinders it would have been a tremendous task to remove all moving rods to comply with the 'dead haul' rules applied by B.R. But thanks also to Bill Harvey of Norwich and Geoff Sands, it covered the four hundred miles from Ayr without a hitch and with no suggestion of over-heated axle boxes or bearings which were the bugbear of dead-hauling disused locomotives. For lack of a suitable and safe siding nearer to Bressingham, it was hauled via Ely and Thetford to Norwich and placed ready for drawing back to Thetford once 'Royal Scot' had been unloaded at Bressingham.

A telephone call woke us past midnight. A solicitor was on his way, I was told, with an Injunction to prevent 'Duchess of Sutherland' being moved from Norwich. Later, another call announced that the Injunction had indeed been fixed to her cabside, and the thought of a legal battle ahead prevented any more sleep that night. But seven days later, at Leeds Assize Courts, the case was dismissed with costs because when it came to the crunch the plaintiffs simply had no case. It was Butlins' property and they had the right to decide where it should go. Later, when the truth emerged, one could only feel sympathy for the Plaintiffs. They were a young band of enthusiasts who had been duped by someone else who should have known better than to incite them to take legal action in the belief that this would divert 'Duchess of Sutherland' to their own preservation centre.

The headaches over the Butlins locomotives were pretty constant during that period. But when the hubbub died down, as it did once the four forlorn, rusty engines were eased down at Bressingham, we saw that, exciting as the new acquisitions were, there would be plenty more headaches to come, before they were fully brought back to active life. It would be a long, tedious job, and until they were scraped down and dismantled there was no real telling what we had let ourselves in for.

For Geoff Sands their arrival was of very special interest. Both 'Royal Scot' and the 'Duchess' had been prepared at Crewe works after being purchased by Sir Billy Butlin during his period there as Loco. Shed Master. They were then, in 1962, repainted in maroon L.M.S. livery, but were never intended to run again under their own steam. The same applied to the L.S.W. Shunter, originally from Southampton Ocean Terminal, and the 1875 Martello, an 0–6–0 Brighton Terrier which had stood with 'Royal Scot' at Ayr. But salty sea air had eaten into paintwork after nine years' standing, and it was easy to note which side of each loco. had been facing seawards. As Geoff looked them over lovingly, it was also easy to imagine his thoughts being, 'All right, my beauties, give me time and I'll give you back your pride.'

The fact was that we could but take our time in restoring these four fine engines. We'd been lucky to have them allocated and they deserved the best we could give them. Some replacement parts, such as boiler tubes, took about nine months to be made and delivered, and some parts might have to be specially cast or forged. Don Hubbard had just spent weeks cramped up beneath the rusty tender of 'Oliver Cromwell' (not included in the B.R. overhaul) scraping every nook and cranny before applying preservative and paint. He too looked over the new arrivals and probably also said to himself, 'I'll see it never needs doing again in my lifetime.'

With the 1971 Open Season about to begin, there would be very little time in which to make the new arrivals more respectable. Twice weekly running meant four days spent on preparation, cleaning and maintenance. 'Oliver Cromwell' alone took 10–12 man-hours to bring back smart running conditions after each time it was steamed. My own commitments by now had of necessity brought me more into the garden and its plants which I'd tended to neglect for the past year or two in favour of engines. All I could do for 'Royal Scot' and the 'Duchess' was to put a rough notice-board on each for visitors to read, describing briefly the engines' history and our intentions for them. But the estimate I gave of the 'Scot' costing £3,000 to restore and the 'Duchess' £5,000 showed far more optimism than accuracy on my part, as did the estimation of the time it would take.

By this time my efforts, begun in 1967, to put matters on to a proper footing by forming the venture into a Trust, appeared to be close to fruition. It had been a worrying and quite expensive process, involving consultations with legal experts. When the latter advised a link-up with the Transport Trust as the best means of ensuring continuity beyond my lifetime, I imagined we would come to a swift and satisfactory agreement. Until 1967 I had no idea that by making donations to charitable causes I was, strictly speaking, on the wrong side of the taxation laws. Having carefully kept all gate receipts in a separate bank account for distribution at the end of each season, I believed these donations to be perfectly in order, provided all other receipts were taken into account for tax. But the first professional man I asked how to set about forming a Trust expressed doubts. They said that it might help if two outside signatures were added to mine on cheques paid to charities and some other payments relating to the Museum project, and this was arranged. But these doubts spurred me on to form a Trust, which I imagined would take no more than a few months.

Snags and difficulties, however, appeared round every corner as far as forming an independent Charitable Trust was concerned. It was these which drove me into the arms of the Transport Trust a year or so later. More meetings and consultations followed. Draft agreements were prepared and when finally both the Trust and I were satisfied on major clauses, either the advising Counsel or the Charity Commissioners turned them down. Maybe there would have been no problem had I been willing to relinquish control at this end and hand over all assets to the Transport Trust. But though the latter were fully in agreement with my desire to keep control and finish the task I'd begun, there were still legal snags affecting the vital aspect of finance. As a Museum project it could not develop and be properly maintained if there were any doubts about money. We had to have guarantees of first claim on gate receipts for restoration and maintenance of engines, for it was obvious now that outside charities could no longer be supported, when such costs as restoring the Butlins locomotives had to be met from this one and only source of income.

It was all very tricky indeed. The Transport Trust were not only willing but anxious to co-operate. As a young and little-known body, to have Bressingham under its wing, as it were, would be a considerable asset. They saw Bressingham as a kind of shop window, and because by 1970–71 we were all convinced that legal difficulties would be quickly ironed out, it was thought quite safe and above-board to make donations to the Trust in order to cover haulage costs of engines to Bressingham as well as heavy restoration expenses such as boiler tubes. The keynote for all these efforts was to keep all things above-board and get on the right side of the law as quickly as possible, segregating the Steam Museum as a non-commercial venture, in no way dependent on or connected with the Nursery and farm. It would be as wrong for the latter to subsidize the Museum as it would be for any of the gate proceeds to be used for a non-charitable purpose. The idea of opening to the public had never been conceived or carried out as a means of making profits, but purely as an opportunity of raising money for charity.

In law, ignorance is, however, no excuse, and I was doing my level best to put matters right. I was not entirely happy with the final draft of an Agreement with the Transport Trust, for it seemed nothing could be done to prevent the Council of the Trust having the final say in controlling the Bressingham finances. The existing laws allowed no other interpretation, if by handing over my privately owned engines crippling capital gains tax and estate duty were to be avoided. The Trust was a registered charity and my overgrown hobby was not. Even though I'd never thought of it in money terms, of making profits, the dice seemed to be loaded against my keeping control. It was galling to find there were penalties for wanting to continue working as custodian of what had become a fully-fledged, live Steam Museum for the public to enjoy and was not to be dispersed at my death. It was already a collection of historical importance, but as relics, my own engines were now greatly enhanced in value, and though I wanted to give them, as it were, to the nation, till now the law had made it extremely difficult for me to do so without losing control or being liable for tax in some shape or form.

As 1971 came to an end, the link-up with the Transport Trust seemed the only way out. I had misgivings, but managed to swallow them for most of the time whilst the final draft Agreement went to the Charity Commissioners for approval. This Body, notoriously slow to reach decisions, had not been unsympathetic. But no doubt they had to be meticulous, even if by now it was pretty certain that approval would be given. On the strength of this, and at the request of the Transport Trust, there was to be an official 'handover' ceremony at Bressingham during 1972. The date, June 24th, was fixed. Invitations were sent out and the press was advised. As a ceremony it did not appeal to me, and at the back of my mind was the niggling fear that someone, some day, perhaps years ahead, might have cause to regret it. Increasing gate receipts proved that we could be financially self-supporting. As a Museum the collection was comprehensive and alive, large enough to merit being a Trust in its own right—if only the way had been clear to make it so. If only the Law had been less restrictive on private museums. . . .

But when June 24th came, and I handed over documents to General Lonsdale, representing the Transport Trust, only a very few knew that those documents were still incomplete. The Trust's Solicitor had still to make adjustments to one or two clauses, and until this was done the Agreement and Deed of Gift, as well as a

21-year lease on the Museum property, could not be signed. At that time I had no other thoughts than of signing within a week or two. So far as I and my family were concerned, the best arrangements possible in the circumstances had been concluded, and as a going concern the Museum was being taken over by the Transport Trust as parent Body. But with the ceremony over, the weeks slipped by and still the Trust's solicitor held back the unsigned documents. There were only minor points, I was told. He'd try not to keep me dangling much longer, and the Trust were equally anxious to have it all buttoned up.

But it never was. During the latter part of the summer news filtered through to me that there had been some relaxation in the Law relating to private museums with collections of historical and educational importance. On checking up I found this was indeed the case, and after much heart-searching I decided that we must take advantage of it and go back to the original idea of forming a Charitable Trust of our own. The Transport Trust would be disappointed, but I hoped they would not consider I had let them down. The pressure to be independent, as well as the advantages, were too great to resist, and back I went to square one after five years of anxiety and effort, to say nothing of expense in legal fees.

Among the locomotives which came to Bressingham with haulage costs paid for by the Transport Trust with funds we had provided from gate receipts, was a Norwegian 2–6–0. It had been purchased by Gerald Pagano, and I was willing enough to provide him with a place to keep it, in the hope that once here, it would stay. Gerald worked like a Trojan, mostly alone, scraping down and repainting his engine—a handsome thing, especially when, in late 1970, it emerged into the open in shining apple-green livery.

Having the right to use it on open days, Gerald also felt a right to steam and drive it, but after a few steamings in 1971, 'King Haakon VII', as he named it, proved to be something of an embarrassment. For safety's sake we had to lay down rules, and one of these was that only qualified, experienced drivers should be at the controls when visitors were around. Gerald was not in that category, and when Geoff Sands, as C.M.E., had to remind him of this rule, Gerald decided to have done with us and take 'King Haakon' elsewhere. Sadly, we had to let him go, but Gerald found other steam centres were also restrictive, for in the next three years the locomotive went from here to Quainton Road, then to Ashford and lastly to Loughborough.

It was when Geoff went to Dagenham Docks to see 'King Haakon' safely unloaded that he also saw 'Solomon', an 1877 Manning Wardle Shunting engine. This was on a plinth in the forecourt of the Samuel Williams Dock premises. It had been a faithful servant to this firm, with its extensive docks, for so long, that when it was retired in 1958, it was given a place of honour and roofed over just for sentiment. As a result of a letter I wrote a visit followed and since the engine was deteriorating under its open-sided canopy, the firm agreed that it should come to Bressingham for permanent care once the Managing Director had come to make sure we could provide a suitable and secure home.

The next arrival had also spent its working life in a dock area. A well-wisher gave us the tip that the famous Doxford and Sunderland firm were dispensing with their crane engines, and that these would shortly be for sale. This, I thought, was an opportunity for Jim Price, who had bought the Garratt 'William Francis' for the Museum. Having three young sons, it appealed to him to acquire an engine for each of them—at least to be in their name, though to be restored and kept at Bressingham. Being first in the field, we were able to have the pick, and in due course 'Millfield' arrived on a low loader all the way from Sunderland. It was rough, but sound, having only just come from work and having been built by Robert Stephenson and Hawthorn as recently as 1942. The forward projecting crane was used for ship-building and repairs and had a 5-ton weight capacity; but this was no ugly monster and as an 0–4–0 weighing 32-tons it might, I thought, be of some practical use.

As a collection of several types of both industrial and passenger locomotives, the Museum was becoming pretty comprehensive by 1972. But there was no freight locomotive. One had been in the offing for some time, but I was not whole-hearted about acquiring it; nor was John Scholes. As Curator of B.R. Historical relics, he now had to consider the Department of Education and Science who were due to take over the National Collection and so relieve B.R. of a responsibility which they no longer wished to shoulder. Hence his hesitation over the J.17, the one freight engine in store at Brighton which, if not required for the proposed new Museum at York to replace that at Clapham, could appropriately come to Bressingham, since it had spent all its working life on East Anglian lines and had been the last working steam engine in the region, based at Norwich. My own reservations were entirely different and quite personal. The 1903 J.17 was not only a plain 0–6–0 with inside cylinders, which showed fewer moving parts, but it was the very type which had been the subject of my nightmares in early childhood.

These considerations were of course pushed aside when a definite offer came for the J.17 to make the collection more complete. But when it arrived, an examination showed that much work would be needed before my old memories of seeing and hearing it in motion under steam could be revived. After being withdrawn from Norwich in 1962, it had gone to a B.R. works for restoration and repainting. But the paintwork had been skimped. Flakes of black were already peeling off to reveal rust beneath, and a few small but vital parts were missing. With priorities now centred on 'Royal Scot' and 'Duchess of Sutherland', the much less spectacular J.17 would have to wait its turn.

'Royal Scot' was in fact nearing completion by this time. The cost, in both time and money, had trebled that of my estimate, made soon after it arrived. Every square inch of its metal work, inside and out, beneath and above, had been scraped free of rust and painted several times. The motion had all been taken apart, including valves, pistons, reversing and braking gear, injectors and cab fittings, and not replaced until clean, bright and in good working order. A complete new set of small bore tubes had been fitted, but the large tubes, in which the superheater elements tubes ran, were a tedious task. They had to come out because until the boiler itself was empty of tubes a thorough test and cleansing could not be made. There was the fear, too, that they would need renewing, but being screwed in at the firebox end they were so tight that they had to be cut out—all of them 4½in. in diameter.

But there was another fear. It had been planted with us by an ex-B.R. fitter at Nottingham Shed, where 'Royal Scot' was last in use in 1962. He told us that the reason for its being withdrawn from service was a fracture in the main steam J pipe which took the superheated steam from the boiler to the dome and so to the cylinders. If this were true, and we were assured it was, then the pipe had to be renewed or put to rights. Tests of several kinds, including electronic, were made with the boiler cleared out, but no fracture showed up. It was puzzling and very worrying but in the end Geoff decided it could only have been a leaky joint. Yet there was the lingering fear that, when all other boiler work had been renewed or replaced, then the hidden fault would again become evident when under full steam. This would have been disastrous, because the large element flues had all been re-threaded on renewed ends, 18in. long, welded on at over £19 each.

But Geoff's hunch was correct. The man from Nottingham could not have known the exact cause of the trouble because no inside examination of the boiler could have been undertaken at the time of withdrawal from service, without disembowelling it as we had done. He had, it seemed, jumped to the conclusion that the J pipe was fractured, for both water and steam tests at extra pressure proved successful. In its splendid maroon livery, lettered and lined out by Bob Rolfe, a local craftsman, 'Royal Scot' went into service on open days that summer, a delight to behold. But it had cost over £10,000 to bring it back once more to life and splendour.

For three seasons 'Oliver Cromwell' had plied up and down the 200-yard track north of the Museum. The curve was quite severe and apart from causing undue wear on the wheels, there was an element of danger to unheeding visitors who might get in the way. With this in mind we decided to make a new running track and the only place for it was the tree-lined lane, known as the Causeway, leading straight down to the Fen and the River Waveney. I was bothered about ballast until a B.R. track expert told me sand would be the best material. He reckoned tree roots would very soon find it and in the process hold it as a firm bed over the unstable peaty soil which lay below. This was a saving, too, because the nursery had a large amount of plunge bed sand which had become too weed-infested for further use.

Dozens of loads of sand were spread out, and after a month's hard labour nearly five hundred yards of track were laid with three sets of points just west of the Museum. These were to enable locomotives from the Museum to gain access to the new line, as well as to half the double-bay Dutch barn which we had to take over for storage of those awaiting restoration. What had been a farmyard had become a railway yard with seven tracks and six sets of points connecting them.

'Royal Scot' was completed just in time to take over Sunday

duty from 'Oliver Cromwell'. 'Thundersley' and the '2500' were used mainly on Thursday openings, by way of a change and because their cab capacity was too small to cope with the extra demand on Sundays. But 'Oliver Cromwell' had been giving trouble. We had replaced several leaky tubes with new ones before the 1972 season began, and then when the season did begin, superheated elements began leaking. As the high-pitched hiss of pin hole leaks were heard, these were drawn when cold and spot welded before steam was raised again. It happened three or four times and though not in the least dangerous, it was disquieting because more expense was indicated. And in the absence of detailed information, we could but conclude that this, the last locomotive to be overhauled by B.R., had not been very thoroughly examined.

Perhaps those responsible had reason to believe a complete overhaul would be a waste of time and money. When the locomotive was taken in to Crewe works late in 1967, the run-down of steam had reached its last headlong rush, and since 70000 was the Britannia class on the preserved list, it was probably thought that No. 70013, 'Oliver Cromwell', would go for scrap within a year or so. At any rate, no new tubes or superheaters were fitted (though at the time redundant stocks of new ones were being sold for scrap for five per cent or less of what it would now cost us to purchase new) for there was no doubt, when a cross section of tubes and superheaters were taken out as a test, that the whole lot would have to be renewed.

This was another expense that would have to wait. Boiler work had begun on 'Duchess of Sutherland', as well as restoration on 'Martello', the little L.B.S.C. Terrier. It was nearing the end of summer when news broke which could have led to our taking on yet another rescue project. 'Flying Scotsman' was languishing on the Pacific Coast of the U.S.A. with no hope of its owner, Alan Pegler, getting it back. He had in fact gone bankrupt with the failure of his venture over there and was being hard pressed. Much as I would have liked to see it join its old rival 'Royal Scot' at Bressingham, the obstacles seemed too formidable to contemplate. But the thought nagged, especially when one or two people suggested that, if I would help to get a 'Save the Scotsman' project in motion, they and many more would back it. Though the more I thought about it the less likely it appeared that such a famous locomotive could or should come to Bressingham and stay there, I was none-the-less keen to do what I could to bring it back to England. A series of telephone calls led to meetings, first with the Official Receiver, and from Alan Pegler I learned how complex was the situation. American creditors were pressing hard as well as those in the U.K., and swift action was demanded. We formed a consortium willing to act as guarantors pending a public appeal for funds, and then after two meetings it became a matter of urgency for someone to go to America to bargain with the creditors for the release of the locomotive for shipment back to the U.K.

Although I supposed I'd succeeded in getting something moving, I was not anxious by now to become too deeply involved. There were some willing guarantors amongst those who had been asked to join the consortium, only one of whom had the financial resources necessary to cope with the 'now-or-never' emergency which now appeared. This was Bill McAlpine and he it was who offered to pay the expenses of an exploratory trip to America by George Hinchcliffe, manager for Alan Pegler's 'Flying Scotsman Enterprises Ltd.' This put an end to the burden of an almost daily series of telephone calls I had had to make as the one to tie in arrangements and consultations, especially as from the first George Hinchcliffe became a key figure in the matter.

George flew to America just after Christmas, and New Year had only just begun when news came that a deal had been possible because Bill McAlpine had been willing and able to put up all the money necessary to buy the locomotive outright. 'Flying Scotsman' arrived back in England within a matter of weeks, and I could not be other than thankful that Bill had stepped in as he did. With an overhaul and repaint undertaken, this most famous of all locomotives was to be for the nation, when a further limited series of nostalgic journeys came to an end.

27
Growing Pains

The first three years of the seventies had brought plenty of excitement as far as standard gauge locomotives were concerned. They had also brought anxieties, but these were, I was fast learning, to be expected. There were compensations, however, especially in the joy it was to see the locomotives as one by one they came back to life. There was also proof in those years, in the steady increase of visitors, that not only were we working on the right lines as a private museum but also that the garden was a kind of museum too. Bressingham had a uniqueness not shared, or likely to be shared, by any other place in the country. As a nurseryman by profession, retired from active control of the business now being run by Robert and Adrian, I was still finding too little time for all I wanted to do. Several weeks in both spring and autumn were from choice as well as necessity spent in caring for my collection of plants in the gardens. But during winter months and for odd days in summer when not driving 'Bronwllyd' on open days, I enjoyed the manual work involved with the narrow gauge railways.

These, too, developed in one way or another during these three years. Having lengthened the Nursery Line to its fullest extent of just over two miles, I found in driving round at least thirty times a week behind 'Bronwllyd's' smoking chimney, that there were still improvements to be made. Some bends were too cramped, and the heavier Orensteon and Koppel had made widening them even more necessary to cope with the demand. This was met by two trains running on Sundays, with one departing as the other arrived back.

When such tasks were completed late in 1970, the notion occurred to me that with an old track bed still in existence on the north margin of the little lake, it should not be very difficult to make a complete circuit and so provide another ride of a good half-mile. Enough rail and sleepers were in stock, but knowing the turns at each end of this long narrow stretch of water would be pretty severe, I reckoned $10\frac{1}{4}$-in. gauge should be workable. It so happened I knew of a locomotive of this gauge in Norfolk, owned by a private collector who had no track on which it could run. An approach was made to him which led to a hiring agreement, since he did not wish to sell his little Atlantic, 4–4–2 'Peter Pan', built as a perfect scale model to the old Great Northern design in 1909—an early Bassett Lowke. Its past history included the conveyance of royalty, for at the 1924 Wembley Exhibition King George V and Queen Mary were amongst its passengers. But it proved to have rather severe limitations at Bressingham. During the winter, I laid sleepers on a track bed I'd formed on the south side of the lake. To achieve this, a high hedge had to be cut back in places and two poplar trees removed. On this side the bank was lower and only two feet or so above the lake water level. It allowed for a quite long strip of marginal ground which I dug and planted up with moisture-loving subjects, just to add colour and interest.

But at each end there were problems. Not only was the turn unavoidably tight, but whichever way the train ran there was a gradient to be climbed—a difference of three to four feet in level between the north and south banks of the narrow-ended lake. It was impossible to ease these curves, but when 'Peter Pan' began trials there was, we thought, power enough to pull the three small trucks with passengers aboard since it made no real bones about hauling them empty. I would not have believed a score or so of children and adults would have made so much difference, but even half that number proved too much for efficient running.

Having to limit the load, and the driver sometimes having to nip out and push on the gradient, was no joke. What I'd hoped would be a picturesque ride, able to pay its way, was a flop, and at the end of the season 'Peter Pan' was returned to its owner, none the worse for its stay with us. In its place came 'Prince Charles', a 4–6–0 Royal Scot class built for heavier work. This had been in use on the sea front at Rhyl, and after a few repairs took over for the following season. Curves and gradients were eased a little more but the first high hopes were again whittled down. On the straight, level parts 'Prince Charles' developed both speed and power, but not enough to complete the circuit without losing so much steam pressure that the next passengers had to wait for it to build up again. Geoff Sands, Reg Cooper, John Bilton and Philip Gray had all taken turns as driver. All tried their hardest to succeed in hauling pay loads, but 'Prince Charles' was very little more useful in this respect than 'Peter Pan' had been. The Lakeside Line was still a flop, and so far as I could see would never be a success. Any more powerful locomotive of that gauge would still jib at those tight end turns, and no wider gauge could be considered.

But I did consider 15-in. gauge using either the north or south lakeside track bed as part of the run. The obstacles to this were not only to find a different route beyond, so as to form a circuit, but to find 15-in. gauge rolling stock. Two very successful railways in this gauge were already in existence—the Romney, Hythe and Dymchurch and the Ravenglass in Cumberland. The chances of finding stock seemed remote, but having decided to sell 'Prince Charles' and have done with $10\frac{1}{4}$-in., I could but make enquiries. By the end of that summer of 1972 the only information was that none were for sale and that to build a new one would cost at least £20,000. This was shattering, though I'd heard of a new $10\frac{1}{4}$-in. that year costing over £10,000. 'Prince Charles' had cost under £1,000 and at the time this seemed a lot. Perhaps the fault was mine in not being realistic, or I should never have imagined the Lake as a suitable circuit. Then one Saturday afternoon a casual visitor braved the deterring closed gates and notices. Had he not given me a reasonable explanation, I would have turned Peter Nicholson away, but instead I let him see our narrow gauge locos in which he was specially interested. He did not stay long and just before he left I mentioned fifteen inch. 'None I know of in this country,' he replied, 'but I heard the other day that there's three for sale with coaches somewhere in Germany. I don't know whereabouts but the chap who told me might know. I've got his address and phone number at home.'

It was not long after, following a few telephone calls, that an address in Germany was available. A letter typed in German by Sheila who was on the Nursery Staff was sent off, but after waiting three weeks with no response, Sheila made a telephone call. It misfired. The man at the other end had left his position as manager for the owner near Cologne some time before; but at least we were told who he was, though the man believed the locomotives had already been sold to someone in England. Another call was made with hopes at a pretty low ebb, but ten minutes later, I was planning a trip to Cologne. They were still for sale, at what appeared to be a bargain price for three locomotives and 29 coaches.

I set off for Cologne a week or so later convinced that I was on to a good thing, but I found on meeting the owner that there had been a misunderstanding. The price quoted had been for only two locomotives and 19 coaches. He wished to keep the remainder for sentimental reasons, and I could no longer see this kind of deal as a bargain, knowing that the cost would stretch our resources to danger point. Restoration commitments and other looming expenses were such that I had to ask for a few hours grace in which to decide. I walked miles that evening, and slept scarcely at all that night. Morning came and I knew the plunge had to be taken if only because I could not bear the thought of going back empty-handed. The sight of those handsome Krupp Pacifics in their dim shed with all those splendidly built coaches had gripped me so much that I convinced myself they would pay for themselves at Bressingham, given time.

The deal was arranged that morning in front of a Cologne lawyer. Transport was handed over to a large firm of forwarding agents who were asked to inspect and quote. If the figure of £1,750 for transportation which they subsequently asked seemed rather steep, I had to accept for lack of other means and practical offers, for the sooner the stock arrived at Bressingham, the better. As it happened there were unexplained delays and then, with Christmas only a few days away, three Dutch juggernauts arrived early one morning. Not knowing the kind of vehicle to be used we hurriedly drew out and piled up the 19 coaches close to the Museum, using our own ancient crane dragline. Measuring the height of the lorry floors from the ground, we built a ramp down to a short piece of 15-in. track we'd previously prepared, but it had taken the juggernaut an hour or two over muddy ground to reach the spot. Inside were the two locos—'Männertreu' and 'Rosenkavalier', and by now it was nearly dark and raining again.

Without the tender, each loco weighed nearly seven tons and it was very tricky indeed to entice the first bogey wheels from the timber racks on which they had travelled inside the lorry on to our ramp, built up with sleepers nearly five feet high to match the lorry floor. Having jacked up the first loco and replaced timber with rail to form a continuous run, the next problem was how to check a headlong rush once the weight was on the ramp, as neither a winch nor snatch blocks could be used. The nine or ten of us unloading could not have held it. Then Robert suggested using the compressor and so applying the engine's own air brakes. A good idea. And once connections and tests were made we inched it out from the level lorry to the steeply inclined ramp.

By now it was pitch dark and still raining. The braking system was fine, but 'gently does it' we said as we stood back to watch in the gleam of torches and hurricane lanterns. When the loco had all its weight on the ramp, suddenly it plunged forward, unchecked, lurching and almost disappearing into the darkness. There was a sickening thud as it pitched over on to its side, unable to take a curve we'd had to make in the track to keep clear of the Lake. Sighs of relief came on two counts. One was that the engine appeared to be undamaged, thanks to muddy ground; the other for its having pitched over to the right. Had it gone to the left it would have plunged into the lake, for at this point the bank was pretty steep and the water quite deep.

The unsuspected dip in the ramp which had made the brakes ineffective was put right, but it was midnight before the second locomotive was unloaded. Next morning I found Geoff and the rest of our helpers in the process of placing the casualty upright again. It did not take long. From remarks made the evening before, I gathered that the newcomers were very welcome. Now, in daylight, they were the only topic of conversation. Remarks made about their appearance and how well both they and the coaches were built confirmed my own opinions in no uncertain way. All were delighted, but with a note of impatience to see them running again. But there was work to do before that day came. The locos were shabbily painted in Prussian blue, and some of the teak wood coaches needed repair. It was many years since they last ran, and though they'd been overhauled a six-year storage since then had taken some toll.

Over Christmas, I put down marker sticks for a route. Leaving the south bank of the Lake, it would have to run beside a large ditch and, beyond that, a smaller one would have to be filled in. Then, crossing a Nursery roadway, on turning south, it would also have to cross the existing Nursery Line on the level, and this would be a tricky job for a welder. Then, passing through a group of oak and birch trees, it would go due south, on the line of a sprawling hedge of brambles, due for cutting back anyway, before reaching the Waveney River by-pass. A new bridge would have to be constructed, though one had already been planned as a means of access to additional land acquired the year before. More hedge trimming would be needed beyond, and then a turn to the west along the back of the old Waveney River.

After Christmas, Ben, Arthur and I set about the task with a will. My job was clearing the way, and with the help of George Orford and the digger tractor, I made the formation and laid the sleepers. Ben and Arthur followed with track laying, and by late February, when the seasonal garden call came for me, they were left with less than half a mile more to lay. It was ready for the open season, but only just, and because the full circuit could not possibly be completed, both we and the visitors had to be content with running half-way—about a mile out and then back on the same track. No track laying we have ever done on this often soft ground has been right first time and this was no exception. Settling embankments and over-tight curves called for adjustments after the first few runs, but it was a heartening sight to see the train, with one engine heading it out and the other at the rear to haul it on the return.

Much time and thought had been given to deciding how the circuit should be completed. It could but take virtually the same route for the last three-quarters of a mile of the existing 2-ft. gauge Woodland Railway. To merge the two by means of a third rail would be to invite trouble, not only with points but also with stoppages from the signals which would be needed. In any case the Krupps could not take such sharp curves as the 0–4–0 Hunslet 'Gwynedd', which with 'Maid Marian' had powered this line. As a ride, the Woodland was not so very popular, and with great reluctance I decided to discontinue it, with a grim memory of the sweat and toil I myself had put into making the return loop which would have to be abandoned completely and taken up. It was not the first occasion that, either through trial and error or some unforeseen development, a task had to be undone and begun afresh. In this case, due to other pressures, it was not until late in the summer of 1973 that time was made to continue the Waveney Valley Line, as we decided to call the new route. Ben and Arthur worked their way through two poplar plantations which a neighbour had kindly given us leave to skirt, keeping to the bank of the old Waveney River, the Norfolk–Suffolk boundary. From there, entering our own wood of pine, oak and rhododendron, it was a case of taking up the Woodland Railway track after cutting out the worst of the curves and relaying it to the 15-in. gauge. It was not until late November, with garden tasks ended, that I could join in the work, but by the time spring called me back the circuit of 2¼ miles was complete, ready for the two open days at Easter. Painted now in deep Brunswick green, each with a nine-coach train, 'Männertreu' and 'Rosenkavalier' were splendid in performance as well as appearance. And the number of visitors who braved a cold east wind at Easter to take the journey was a pointer to some right decisions having been made after all.

It had become increasingly obvious that these narrow gauge railways would in fact be lifelines for the future. We were lucky to have interesting routes for passenger traffic and what profit ensued would certainly be needed to restore and maintain the collection of other locomotives and traction engines which brought no return when steamed just to keep them alive. I'd been rather too timid at first, though not too late to acquire narrow gauge locos before supplies dried up and prices went sky-high. But those I'd bought were old and the time would come when expensive new fireboxes and boilers would be needed. Reserve engines were therefore vital. So once the narrow gauge had proved their worth, I was always on the look out for one or two more. The first addition to the 2-ft. gauge stable was the O. & K. I'd bought in Bremen, which proved to be too heavy. I'd had no luck at all in trying to obtain one or two more from Northern Spain in 1968–9, but in 1971 another was offered from Germany. It was an 0–4–0 Schwarzkopff which I bought unseen, with assurances that it was in good order. But this also took several weeks to bring back into running condition, many worn bearings having to be renewed and new tubes fitted as well. The weight of this had been given as 6½ tons, but our guess was that it was nearer 8 tons and of very robust Teutonic appearance.

On a searching trip to Vienna in 1970 I'd seen a fascinating little Krauss. The owner, a secondhand machine-tool merchant, offered to sell for the equivalent of £400. But I could not pay a deposit due to currency regulations and when I wrote on my return offering to send money against a contract to legalize a sale, there was no reply. I wrote altogether four times in as many months, and forwarding agents made personal contact for me, but nothing came of this except vague promises which were never fulfilled.

By 1971 'George Sholto' proved to need a new firebox and had to come out of service. Two quotations from amongst the very few firms able and willing to undertake such boiler work were frightening. Then a man called who offered to make a good job of it at his Dartford workshop. He seemed to know what he was talking about and the price quoted was well below the other two. The engine was stripped and the boiler was taken away with the promise that it would be back, tested and ready for re-assembly, by March. April came and both letters and phone calls were unanswered. The firm—two brothers—had left the district, so the police reported, and there was no trace of 'George Sholto's' boiler at their vacated premises. A few weeks later, word came from an interested stranger that he'd seen the boiler in a breaker's yard fifty miles away in Surrey. Geoff went to inspect and found that some work had been done on the firebox, but there was no news to be had as to where the brothers were, only a report that the job had been farmed out to someone else. Eventually, again with the help of police, 'George Sholto's' boiler was returned, and foolishly, perhaps, I paid the man who said he was unable to finish the firebox because of illness £300 for what he'd done. Months later it went to a reputable boilersmith works in Durham, and after inspection by a Lloyds official, the work already done was condemned. The whole job had to begin again, with the addition of a new boiler to go with the firebox and smoke box at a cost of £2,000. I was still searching during 1973 when the chance came to exchange the big O. & K. for a smaller one of the same make. A company hoping to run another narrow gauge railway in North Wales needed more power than the smaller 'Eigiau' they possessed. Although both locos needed attention, a level swop was made. It was tricky, because by the end of the year 1973 'Bronwllyd' too had to come out of service. This 0–6–0 Hudswell Clarke had been my regular engine to drive since 1969. In making an average 800–900 trips on the Nursery Line every year since then, I'd got to know her pretty well. She had never let me down, and on one occasion when 'Brunhilde' the Schwarzkopff failed while sharing duty on this line as the second train, 'Bronwllyd' pushed it with a full load three parts of the way round, as well as hauling her own load—a total of close on two hundred passengers. But the time had come for a complete overhaul, including new tyres, bearings and firebox. We have five locos altogether of this gauge, yet in spite of discontinuing the Woodland Line only two were fully serviceable for 1974. Our 'Maid Marian' friends, by the way, had decided in 1972 to take their little loco back to North Wales to be nearer home.

It was largely due to the vast amount of work on 'Duchess of Sutherland' that overhauls were put in arrears on the 2-ft. gauge. Firebox repairs and tubing on the 'Duchess' had begun in 1972. All kinds of snags were encountered, often because spare parts and replacements were virtually non-existent. By way of example, one of the small tubes needed a die nut of an odd size to ensure a good fit in the firebox end—the box being made of copper. Enquiries were sent all over the place, over a period of months, but nothing of this size was found. In the end, one had to be specially made, causing further delay and a shock for me on receiving a bill for £47 for a nut of about 4 oz. weight of steel.

The task of restoring the 'Duchess' having been started, it had to proceed regardless. Nothing must be skimped, but we were persuaded that both time and expense could be saved by having the 'Duchess' sandblasted instead of hand scraped and rust, old paint and grime chipped away. This, we were told, would scour down to the bare metal every nook and cranny, especially the underparts of the tender and cab, though it was understood that the motion and cylinders would need protection. All the adjoining locos in the shed also needed protecting from the 800 p.s.i. blast, and 6233 had therefore to be covered in a huge cocoon of heavy polythene. Ten tons of special 'sand' or shot arrived in the late summer of 1972, and for a week the roar of the compressor was heard. At last it was stilled, and clearing the mess, along with cleaning down other engines which were coated with dust, took several men several days. In the process it was discovered that many a nook and cranny had escaped the blast and still had to be scraped by hand. Poor Don, who'd had so much of this kind of work beneath the engines and tenders, was dismayed at the thought of much still to do, in cramped, dark and dirty conditions. The blasting job, such as it was, had cost over £700, and in paying up came the resolve never again to apply this hit-and-miss method, if we ever had another loco in such a derelict condition to restore.

A lesson I learned, as Don, and one or two more, slowly brought hope and pride in place of frustration and near despair, was that until an engine has been dismembered and scraped as it were down to the bone, one simply could not tell what was involved. Geoff and his helpers took over the more mechanical side, removing the motion and all other working parts to ensure they would work again. These included a steam-powered coal pusher, operated by the fireman to keep coal in the tender within reach of his shovel.

It was not until autumn of 1973 came, with the open season at an end, that the next few months could be devoted to the final stages of the 'Duchess'. Having at last satisfied himself that the metal would take preservative and paint, Don went under and round four times more, until the final coat of paint was applied. The tender, needing some new heavy sheet steel replacement where thin, received far more attention than the cab and boiler barrel casing. Several undercoats were applied and each rubbed down with 'wet and dry' emery till fingers were almost raw—and till finally Bob Rolfe, who was to apply the last top coat, lettering and lining out, was satisfied that surfaces were as good as they could be made.

28

Postscript

Having researched every detail concerning the livery of No. 6233, Bob Rolfe applied the exact shade of L.M.S. maroon paint as the final coat, and lined it out with meticulous expertise, using pure gold leaf where this had originally been used. The cost of this final painting and gilding was well over £500, but since the total restoration cost to us had already passed £16,000, it would have been inexcusable to skimp anything now. On 30th May 1974, the 'Duchess' emerged and her fire was lit. She was a sight for sore eyes, and this being our first open Thursday of the season she went into service for footplate rides. But not for long. As so often happens with a fresh steaming after restoration, full circulation and pressure dislodged odd particles of shale in the boiler, and one of these settled beneath the clack box valve, thus preventing the injectors from functioning and allowing steam to escape. It was galling and it happened again on the following Sunday. Even much older, less important engines had a master valve which would have avoided the need to draw the fire, but the 'Duchess' did not possess one, nor did this class—often referred to as the Rolls Royce of express locomotives—possess a drop grate system. Still-burning coals had to be taken out very awkwardly with a shovel with an iron shaft nine feet long.

Relatively minor faults can sometimes cause a crisis when they occur during open periods. The visitors may be fascinated or amused as we struggle to resume running routine, but to us any such incident is anything but funny. From experience we have learned that the pride and keyed-up expectancy of bringing an engine back to life is liable to be offset by the unexpected snag. But since we do our best to make tests to avoid the chagrin we feel when failures occur if the public are present, such incidents are rare.

'Duchess of Sutherland' now takes pride of place in terms of size and splendour in the Bressingham collection. As the only active example of its class in the country, we are entitled to feel some pride in having restored it, knowing that nothing has been skimped. This knowledge, far from enittling us to slacken off, makes us more fully aware of what more remains to be done. With both the 'Duchess' and 'Royal Scot' a standard has been set which must now remain as fixed, a standard we must do our level best to maintain with all the engines under our care. It will take some doing, but there can be no lesser aim than this, even if it takes years to accomplish fully.

To bring and maintain so many engines of many types up to such a standard cannot from our present standpoint be our single aim. As a Museum there is still scope for development, which we think should come as soon as funds will allow. The main Museum building is already too small to hold all the locomotives, and most of the traction engines have to be left outside under tilts. To house these, and various smaller static exhibits, a new exhibition hall is planned for the near future. Once erected, the larger stationary engines can at last be installed instead of cluttering up the yard and odd corners. Some have been around for several years, partly because locomotives had to come first, but chiefly because I could not see far enough ahead to decide where they could be placed with any certainty of their being able to stay there.

A change is needed also in the location of the workshop area. Repair and restoration work should be carried out in a building designed or specially adapted for the purpose, with stores and machine tools conveniently to hand and an inspection pit for locomotives. This too is a high priority.

Space is fast becoming a limiting factor when it comes to considering more buildings. We have a place in mind for a genuine Great Eastern signal box as an outside exhibit. The shell still stands at Oakington—where I lived almost in sight of it for several years. Before long this should be moved, to join up with its interior equipment already here. Its presenter is Mr. Michael Sharman, of the Cambridge University Engineering Department.

It could be said that thirty-eight engines—those on wheels not counting stationaries—are enough to care for. But I would dearly like to make it up to forty, not for the sake of a rounded figure, but to include two outstanding examples of the larger foreign locomotives—a German 2–10–0 and a French 2–8–2. I am convinced that these two would enhance the collection more than any other type still available for preservation. They would, I believe, appeal to visitors as very successful developments made abroad which were not pursued to any extent in Britain, as an evolvement of the steam locomotive which, in spite of being invented in England, diverged as soon as foreigners set about embodying it with their own ideas. To see these two here, cheek by jowl with the range of traditional British designs, especially when in steam and motion, would be a fascinating study for visitors, young and old.

It will not be for want of trying on my part if by 1975 these additions fail to appear as part of the Bressingham scene. The title of the last chapter of the 1970 edition of *Steam Engines at Bressingham* was 'The Unfinished Ending'. In scanning the last page or two it would appear that these, too, could carry such a title. But in my own mind there is now a conviction which was not there in 1970. It is that if or when these projects outlined here come to fruition, it will be time to say 'enough is enough'—and mean it. In reminding myself that I am no longer in a position to take decisions on my own I fancy my colleagues on the Board will be of the same opinion.

For as long as health and the zest for activity continues, I believe rewards for me will still lie, not in any smug pride of achievement, nor in material gain, but in quite different, often little, bonuses. I do not doubt that while I live there will still be a delight in seeing engines in steam and plants in growth. But amongst other pleasures for me is the expression of wonderment on the face of a child gazing at a big locomotive, and the waving of passengers as one of our little trains comes into view of another on a different line. It does my heart good to hear some old countryman say we've done a good job in keeping steam alive, or townspeople remarking on the interesting variety, and what good value they've had from their visit to Bressingham. What is said about the steam side is also said about the garden, and in spite of the near-certainty that perfection in both may lie beyond my reach and power to achieve, it is such remarks or unspoken gestures that come as reward for labours and anxieties and as stimulant for continued effort.

SUMMER 1975

Many months have passed since the foregoing was written. For reasons beyond the publishers' control the publishing date had to be postponed, and this makes it possible to add that some of the 1974 objectives have been achieved. The new workshop with its 60-ft. long inspection pit is almost ready to receive the next locomotive due for complete restoration. As usual, the work was anything but straightforward. The wettest winter for years turned the floor, already below outside level, to a quagmire, and taking out another three feet in the centre for a pit entailed almost continuous use of a pump. Sides caved in, and concreting was a slow, costly job. But it was done, and a deep drain was laid to keep the pit free of water.

The new 'running' shed, 38-ft. by 105-ft. has been built. In this there are two tracks on which locomotives generally in use on open days will be stored. This and the new workshop will keep the main museum building, and its occupants, cleaner and will serve to house some of the traction engines during winter. The foreign locomotives will also use the new running shed for winter storage because they are too tall and wide to pass through the doors of the main Museum building, in spite of being standard gauge. The French 141R No. 73 arrived in March, after a long process of negotiation and careful planning, much of which was undertaken on the initiative of Patrick Evendon who, with his connections with France and his enthusiasm, solved many a problem for us. He not only took Geoff Sands to Reims and Metz last October to inspect the few remaining locos still available but also, when all arrangements were completed in late February, took Roger Garnham and Philip Gray to the Sarreguemines depot near Metz so that they could act as guards when No. 73 was hauled the 200 miles to Dunkirk for shipment.

When safely unloaded from Sunters' big transporter which carried this imposing monster from Harwich, we decided not to attempt to clear the accumulated grime until later in the year. It was thick enough to prevent rusting, but in due course every inch will be scraped clean and by 1976 it should be no longer dingy black but resplendent in its traditional S.N.C.F., green livery. The class, with its relatively unusual 2–8–2 wheel arrangement, was one of the most successful ever built. Its design came from both French and American experience and between 1946 and 1952 no less than 1300 were built by various American and Canadian firms and shipped to France under the Marshall plan. Geoff Sands made no mistake in selecting one in the best possible condition and the S.N.C.F., were most helpful in this direction. I was fascinated by the vast number of cab controls, but scarcely any carried indications of the purpose they served. Anxious to see it come back to life, it was not long before I was piling wood into its 56 sq. ft. firebox, and having raised 150 lb. of steam Geoff Sands found out by trial and error what each valve and handle controlled, and tagged them for future reference. Everything worked, including the automatic stoker for which there were six valves enabling a fireman literally to blow coal by steam into whichever part of the grate area was in need of fuel.

January 1976. The French 141R was steamed several times during the 1975 summer, to present a challenge to our B.R., drivers, an awesome sight to many visitors and a drain on our coal stocks.

Excitement suddenly erupted during June with the news that at last a ship was willing to bring the German built Class 52 (2–10–0) from Sandnes in Norway. For two years one attempt after another to find the safest cheapest means of transport had come to nothing. And now the chance had come, the price of £8,500, just for the sea passage, was more than we could afford. I telephoned the Agents in Norway and told them so. 'What will you pay?' I was asked, and replied 'Not more than £5,000'. 'O.K., we bring it'. A week later the 'Stokksund' arrived in Felixstowe, with Peter Sunter's transporter there to meet it. No. 5856 (the N.S.B. numbering) was rusty, but having been sealed up in a tunnel for fifteen years this was not surprising. Having had the pleasant surprise that it came to life in steam, with no apparent faults, we decided this, rather than the grime protected French 141, must take priority for Don's ministrations to scrape down and repaint.

With the first phase of the Exhibition Hall (of nearly 7,000 sq. ft.) finished, Running Shed and Workshop in use, we must now wait for funds with which to embark on the second phase of the former. This will be costly, but it should be our last big development outlay, but not the last of our labours with over forty engines to keep alive.

APPENDIX

NARROW GAUGE LOCOMOTIVES

PRINCESS

$9\frac{1}{2}$in. gauge Pacific (1947) 4–6–2. Two-cylinder with Walschaert's gear. Though not a scale model this was built roughly on the lines of the L.M.S. 4–6–2 Pacific of the Princess Class by a firm near London in 1947, specially for passenger carrying, with under-scale wheels. It weighs about 1 ton, and has been in service on the 'Garden Railway' since 1965.

MÄNNERTREU AND ROSENKAVALIER

These are identical 4–6–2 (Pacific) locomotives, built by Krupps of Essen in 1937 for use in an exhibition park at Dusseldorf. They also did similar service at Munich (1953) and later at Cologne, after which they were purchased for an amusement park beside the Rhine, but though overhauled for this purpose, were never used. The basic design was by the late Henry Greenly, with some modifications, and for their size are extremely powerful and strongly built, each weighing nearly 8 tons. They arrived at Bressingham late in 1972, and a new track was only half completed by the 1973 open season. The circuit of $2\frac{1}{2}$ miles was completed ready for 1974, using 9 or 10 of the original German coaches to each locomotive. Unfortunately, the new circuit involved discontinuing the 2ft. 'Woodland Line' for the time being, since this new 'Waveney Line' has to use the same track for part of the route.

GWYNEDD

1ft. 11in. gauge Hunslet No. 316 (1883) 0–4–0. This was completely overhauled at Bressingham, being in need of substantial repair to the boiler and firebox. Ex-Penrhvn Slate Quarries, North Wales.

George Sholto

1ft. 11in. gauge Hunslet No. 994 (1909) 0–4–0 saddletank, weight 7 tons. Also from Penrhyn, this was the first locomotive to enter regular service on the 'Nursery Railway'. It came into service again in 1974 after having a complete overhaul, with a new boiler and firebox costing well over £2,000.

Bronwllyd

60cm. Hudswell Clarke No. 1643 (1930) 0–6–0 well tank, weight 8 tons. This was purchased from Penrhyn Slate Quarries merely as a scrap chassis. Later the boiler from Kerr Stuart *Stanhope* was acquired, and after a complete rebuild went into service in 1969 on the 'Nursery Line', on which it made about 2,500 circuits over the next four years. It began life on the Guildford By-pass, before being sent to Penrhyn.

Eigiau

60cm. Orenstein & Koppel No. 5668 (1912) 0–4–0 well tank. One of the few foreign-built locomotives used in England, this also belonged to the Penrhyn Quarries but was never used after being restored by a Midlands enthusiast.

Brunhilde

60cm. Schwarzkopff No. 9124 (1927) 0–4–0 well tank. This was imported from Hamburg in 1971 where it was used in a copper smelting works. Weight 8½ tons.

STANDARD GAUGE LOCOMOTIVES (4′ 8½″)

Martello

Ex-London, Brighton & South Coast Railway 0–6–0 side tank, built by Stroudley in 1875, weight 28 tons. These 'maids of all work' were known as 'Brighton Terriers' and were in service for 50 years or more. On permanent loan from Messrs. Butlins' Camps, *Martello* was at Ayr before coming to Bressingham, where it was overhauled and repainted in its original brown livery.

Solomon

Manning Warde (1877) 0–6–0, Works No. 641, weight 28 tons. On loan from Messrs. Samuel Williams Wharf Ltd., Dagenham. Restored to working condition at Bressingham. This had a long and varied service in industry, and was retired for some years on a plinth at Dagenham.

BECKTON BY PRODUCTS NO. 1

Neilson No. 4444 (1892) 0–4–0 side tank, weight 28 tons. Presented on permanent loan by Mr. Peter Bland and Mr. John Wood. This, with Beckton No. 25 was low built so as to clear retorts etc.

BECKTON NO. 25

Neilson No. 5087 (1897) 0–4–0 saddletank, weight 25 tons. On permanent loan, this spent some years unused on the 'Bluebell Line' after leaving work under the retorts at the Beckton works of the North Thames Gas Board, which presented it to the Industrial Locomotive Society for preservation.

L.S.W. SHUNTER

No. 102 (1903) 0–4–0, weight 40 tons. This spent its working life on Southampton Docks, and is one of the most powerful 0–4–0 locomotives ever built. Since 1962 it has kept company with *Royal Scot* at Messrs. Butlins' Holiday Camp at Skegness.

THUNDERSLEY

Stephenson No. 3367 (1909) 4–4–2, designed by Whitelegg for the London, Tilbury & Southend Railway, weight 70 tons. A fine example of an express tank locomotive, this was selected both for the 1911 Coronation celebrations, and specially restored in 1956 for the L.T.S.R. centenary. Restoration since then was carried out by members of the Norfolk Railway Society, under the direction of Mr. D. W. Harvey, but on arriving at Bressingham re-tubing was found to be necessary.

J.17

Great Eastern Railway, 0–6–0 with tender, inside cylinders, weight with tender 83 tons. This class, built for goods traffic by Holden at Stratford Works, is the only relic of a very powerful class of freight locomotive. It was the last British Rail steam locomotive based in East Anglia, and was withdrawn in 1962 for preservation, and after restoration for a static exhibit in the National Collection, was transferred in 1972 to Bressingham to bring back to working condition.

ROYAL SCOT

4–6–0 3-cylinder, weight 138 tons. Built 1927 North British Locomotive Company, Glasgow. Re-built Stanier design 1950.

The task of completely restoring to good running condition, and repainting to its original livery was long and costly but we are proud to have done this at Bressingham and to have it on active display to visitors.

L.M.S. 2500

L.M.S. Derby B.R. No. 4/2500 (1934) 2–6–4, weight 85 tons. This handsome tank locomotive, with its three cylinders, was built by Sir William Stanier to replace the 'Thundersley' type. Its success led to a large number of similar but two-cylinder locomotives being built for service elsewhere on the L.M.S. Overhauled and re-painted at Bressingham.

WILLIAM FRANCIS

Beyer-Peacock Garratt No. 6841 (1937) 0–4–0 x 0–4–0. The only standard gauge Garratt-type remaining in Britain. With its four cylinders and articulated chassis, it is capable of hauling heavy loads on steep curves and gradients. Weighs 61 tons, but larger locomotives of this type are still used in Spain, Africa and India. Purchased and presented by Mr. J. R. Price, it arrived in rough condition, but has now been fully restored.

MILLFIELD

Crane engine, weight 36 tons, built by Robert Stephenson and Hawthorne in 1942. Works No. 707. Purchased for the Museum by Mr. J. R. Price from Doxford & Sunderland, where it was used on a dock-side locomotive for ship repairing, etc.

DUCHESS OF SUTHERLAND

4–6–2, built 1938 by Sir William Stanier, Crewe Works. This too is on permanent loan from Messrs. Butlins' Holiday Camp at Ayr. As a class these were the largest and most powerful of all British express locomotives, hauling heavy loads on the west coast route to Glasgow. It has four cylinders and weighs 164 tons. Very extensive repairs and repainting (costing over £17,000) were completed at Bressingham early in 1974.

OLIVER CROMWELL

B.R. Crewe No. 70013 (1951) 4–6–2, weight 143 tons. This was the last of the famous post-war 'Britannia' class to be overhauled at Crewe, as well as being the last steam locomotive to haul a British Rail passenger train (11 August 1968). It went new to Norwich in 1951 and ran 700,000 miles on East Anglian lines until sent elsewhere with dieselization in 1961, and altogether ran over 1,000,000 miles. Capable of speeds of over 100 m.p.h. The need for re-tubing prevented this imposing locomotive from being steamed in 1973, but with this completed, it will take turns with others in coming to life again on open days.

TRACTION & ROAD ENGINES

Note: S.C.=single cylinder. S.C.C.=single crank compound D.C.C.=double crank compound.

Manufacturer Works No. Built NHP—Nominal Horse Power

BEULAH, S.C.

Fowler No. 6188 (1890) 6 N.H.P. 9 tons Traction

This was acquired from a Norwich nursery where it had been used for over 20 years for soil sterilizing. Restored at Bressingham.

BERYL, S.C.

Foster No. 2821 (1903) 6 N.H.P. 10 tons Traction

Spent its working life in South Lincolnshire for threshing, etc.

BRENDA, S.C.

Burrell No. 2613 (1903) 6 N.H.P. 9 tons Traction

Used for many years on a North Norfolk estate for saw-bench work, before coming to Bressingham for restoration and preservation.

BLACK PRINCE D.C.C.

Burrell No. 2701 (1904) 8 N.H.P. 18 tons

A once-famous showman's engine in the London area. Last used in 1941/2 on demolishing blitzed buildings, then in a scrapyard until 1962. Top speed 10 m.p.h.

THE PRESIDENT D.C.C.

Burrell No. 2789 (1906) 8 N.H.P. 16 tons Road loco.

This began life as a showman's engine, but was converted for heavy haulage before 1914. Top speed 10 m.p.h.

BERTHA, S.C.C.

Burrell No. 3112 (1909) 7 N.H.P. 10 tons Traction

The first arrival at Bressingham in 1961, and first restored from a derelict state by its late owner Mr. Gerald Dixon of Sudbury.

BESSIE AND CISSIE as a pair, D.C.C.

Fowler	No. 15237	(1907)	18 N.H.P.	22 tons	Ploughing engines

These worked in pairs, pulling a plough or cultivator across the fields. Owned by Messrs. Baldocks, the Haverhill contractors, till 1955.

BOXER, S.C.

Burrell	No. 3962	(1923)	5 N.H.P.	8 tons	Roller

Completely restored at Bressingham, formerly owned by Doran Bros., Thetford.

BUSTER, D.C.C.

Burrell	No. 3993	(1924)	6 N.H.P.	12 tons	Roller

Also first owned by Doran Bros. and restored here. Both were obtained from a scrapyard.

BUNTY, D.C.C.

Garrett	No. 34641	(1924)	6 N.H.P.	5 tons	Traction

The most popular general-purpose engine made by Garratt's of Leiston, Suffolk.

BARKIS, D.C.C.

Robey	No. 42520	(1925)	5 N.H.P.	4 tons	Tandem roller

A light roller of unusual design built in Lincoln.

BRUTUS

Sentinel	No. 8765	(1930)		11 tons	Timber tractor

Given to Robert Bloom as a wedding present, this tractor, with its two engines, was used for timber work and cost £2,500 when new. Speed up to 35 m.p.h.

BOADICEA, D.C.C.

Foden	No. 13708	(1930)	9 N.H.P.	4 tons	Wagon

One of the last steam lorries made by Fodens. Used as such by the owners for some years, then converted to a road tar-sprayer. Speed up to 18 m.p.h.

One of the traction engines is usually driving a threshing drum and straw elevator to show visitors how grain was threshed before the coming of the combine harvester.

PORTABLES (not self-propelled)

BURRELL, S.C.C.

Burrell No. 2363 (1901) 10 N.H.P. 5 tons Portable

This spent all its working life on estate work and in a Suffolk pottery not far from Thetford, and is believed to be the only compound Burrell portable in existence.

YOUNGS (Diss), S.C.

Youngs Not Known (1908) 8 N.H.P. 5 tons Portable

Built in Diss; this is the only one remaining of the 14 the firm produced.

STATIONARY ENGINES (not yet installed)

Vertical, single-cylinder of unknown make, acquired from a local factory. Said to be well over 100 years old.

GUEST & CRAVEN. Single-cylinder, presented by Messrs. Wards' Sawmills of Boxmoor, Herts. This is believed to be of nearly 100 h.p. and at least 80 years old.

MARSHALL. A once-common type for driving works machinery, this is approximately 25 h.p., and is about 60 years old.

SHAND MASON. Fire Engine. This was installed at Crown Point, Sir J. J. Colman's mansion at Whitlingham, Norwich, about 1895. So far as is known it was never seriously used, and has been presented to the museum by the Regional Hospitals Board, as a perfect example of a stationary fire engine.

MERRYWEATHER PORTABLE FIRE ENGINE. This probably dates from pre-1918 and was presented by Mr. J. Mellon of Kelvedon.

The 1935 Leyland Fire Engine at Bressingham has been repainted since being used for a 'Dad's Army' programme. *Bertha* was also featured, threshing for 'Dad's Army' in 1972.

INDEX